Carbon Nanomaterials in Hydrogenation Catalysis

Catalysis Series

Editor-in-chief:
Justin S. J. Hargreaves, *University of Glasgow, UK*

Series editors:
Bert Klein Gebbink, *Utrecht University, The Netherlands*
Jose Rodriguez, *Brookhaven National Laboratory, USA*

Titles in the series:
1: Carbons and Carbon Supported Catalysts in Hydroprocessing
2: Chiral Sulfur Ligands: Asymmetric Catalysis
3: Recent Developments in Asymmetric Organocatalysis
4: Catalysis in the Refining of Fischer–Tropsch Syncrude
5: Organocatalytic Enantioselective Conjugate Addition Reactions: A Powerful Tool for the Stereocontrolled Synthesis of Complex Molecules
6: N-Heterocyclic Carbenes: From Laboratory Curiosities to Efficient Synthetic Tools
7: *P*-Stereogenic Ligands in Enantioselective Catalysis
8: Chemistry of the Morita–Baylis–Hillman Reaction
9: Proton-Coupled Electron Transfer: A Carrefour of Chemical Reactivity Traditions
10: Asymmetric Domino Reactions
11: C–H and C–X Bond Functionalization: Transition Metal Mediation
12: Metal Organic Frameworks as Heterogeneous Catalysts
13: Environmental Catalysis Over Gold-Based Materials
14: Computational Catalysis
15: Catalysis in Ionic Liquids: From Catalyst Synthesis to Application
16: Economic Synthesis of Heterocycles: Zinc, Iron, Copper, Cobalt, Manganese and Nickel Catalysts
17: Metal Nanoparticles for Catalysis: Advances and Applications
18: Heterogeneous Gold Catalysts and Catalysis
19: Conjugated Linoleic Acids and Conjugated Vegetable Oils
20: Enantioselective Multicatalysed Tandem Reactions
21: New Trends in Cross-Coupling: Theory and Applications
22: Atomically-Precise Methods for Synthesis of Solid Catalysts
23: Nanostructured Carbon Materials for Catalysis
24: Heterocycles from Double-Functionalized Arenes: Transition Metal Catalyzed Coupling Reactions
25: Asymmetric Functionalization of C–H Bonds
26: Enantioselective Nickel-catalysed Transformations
27: N-Heterocyclic Carbenes: From Laboratory Curiosities to Efficient Synthetic Tools, 2nd edition
28: Zeolites in Catalysis: Properties and Applications

29: Biocatalysis: An Industrial Perspective
30: Dienamine Catalysis for Organic Synthesis
31: Metal-free Functionalized Carbons in Catalysis: Synthesis, Characterization and Applications
32: Modern Biocatalysis: Advances Towards Synthetic Biological Systems
33: NO_x Trap Catalysts and Technologies: Fundamentals and Industrial Applications
34: Alternative Catalytic Materials: Carbides, Nitrides, Phosphides and Amorphous Boron Alloys
35: Enantioselective Cobalt-catalysed Transformations
36: Noncovalent Interactions in Catalysis
37: Carbon Nanomaterials in Hydrogenation Catalysis

How to obtain future titles on publication:
A standing order plan is available for this series. A standing order will bring delivery of each new volume immediately on publication.

For further information please contact:
Book Sales Department, Royal Society of Chemistry, Thomas Graham House, Science Park, Milton Road, Cambridge, CB4 0WF, UK
Telephone: +44 (0)1223 420066, Fax: +44 (0)1223 420247
Email: booksales@rsc.org
Visit our website at www.rsc.org/books

Carbon Nanomaterials in Hydrogenation Catalysis

By

Edward Furimsky
IMAF Group, Canada
Email: edfurimsky@hotmail.com

ROYAL SOCIETY OF CHEMISTRY

Catalysis Series No. 37

Print ISBN: 978-1-78801-723-7
PDF ISBN: 978-1-78801-755-8
EPUB ISBN: 978-1-78801-780-0
Print: ISSN: 1757-6725
Electronic ISSN: 1757-6733

A catalogue record for this book is available from the British Library

The Royal Society of Chemistry is a charity, registered in England and Wales, Number 207890, and a company incorporated in England by Royal Charter (Registered No. RC000524), registered office: Burlington House, Piccadilly, London W1J 0BA, UK, Telephone: +44 (0) 20 7437 8656.

For further information see our web site at www.rsc.org

Printed in the United Kingdom by CPI Group (UK) Ltd, Croydon, CR0 4YY, UK

Preface

Carbons have been attracting increasing attention as supports in various catalytic applications. It began with the traditional amorphous carbon materials such as activated carbons (ACs) and carbon blacks (CBs). Because of their neutral nature, carbon supports may be suitable for the production of fuels and lubricants *via* hydroprocessing (HPR). Thus, in the case of carbon-supported HPR catalysts, the adverse effects of nitrogen bases present in difficult feeds could be significantly diminished. Consequently, the catalyst life can be extended. Carbon supports maintain stability in the presence of water that is either produced during hydrodeoxygenation (HDO) or is present in reaction streams, as is usually the case with biomass-derived feeds. This was one of the major conclusions of the review on *Carbons and Carbon Supported Catalysts in Hydroprocessing* by the present author, published as part of the Catalysis Series a decade ago.

Gradually, novel carbon materials have been discovered and have entered the world of catalysis science. Among them, the properties of carbon nanotubes (CNTs), carbon nanofibers (CNFs), carbon nanohorns (CNHs), graphene, graphene oxide (GO), reduced graphene oxide (rGO) and fullerenes have been described in detail. The production methods have advanced to the stage that today, carbon nanomaterials are manufactured commercially. Numerous practical applications of these materials have been noted in the scientific literature, particularly those in various fields of catalysis.

As an important field of heterogeneous catalysis, hydrogenation (HYD) has been playing a key role during the conversion of various feeds to fine chemicals required for the production of pharmaceuticals, perfumery products, food additives, *etc.* Apparently, the severity, in terms of temperature and H_2 pressure employed during these applications, is much lower than that encountered in HPR for the production of fuels and lubricants.

Catalysis Series No. 37
Carbon Nanomaterials in Hydrogenation Catalysis
By Edward Furimsky

Published by the Royal Society of Chemistry, www.rsc.org

Because of the severity difference, in this review the former process will be distinguished from HYD.

The database available in the scientific literature indicates remarkable activity and selectivity of some catalysts supported on CNT, CNF, CNH, GO and rGO supports, compared with those supported on ACs and CBs and also on traditional oxidic supports such as Al_2O_3, SiO_2, TiO_2, SiO_2–Al_2O_3 and zeolites. Among active metals, noble metals such as Pt, Pd, Ru and Rh have been used most extensively. To a lesser extent, transition metals such as Ni, Co, Fe, Cu, Mo and W have also been attracting attention. In an effort to improve catalyst performance, bimetallic catalysts consisting either of two noble metals or of one noble metal combined with a transition metal have been developed.

The advances in preparation methodology with the aim of obtaining tailor-made catalysts have been noted. Of particular significance are the methods for modification of the surface structure of carbon nanomaterials. First, the low reactivity and hydrophobicity of pristine carbon nanomaterials can be changed by doping, which involves the replacement of carbon atoms in a hexagonal aromatic ring with heteroatoms such as nitrogen, sulfur, phosphorus and boron. This results in the modification of the electronic properties of the surface. In addition, functionalization can change the acidity and basicity of neutral pristine solids. For example, ranges of acidic and basic strength can be attained by introducing hydroxyl, carboxyl and sulfonic groups and amino groups, respectively. With such supports available, bifunctional catalysts based on functionalized carbon nanomaterials can be prepared. Otherwise, such catalysts could be prepared using acidic supports (*e.g.* silica–alumina, zeolites).

The biofeeds derived from different types of biomass have emerged as an important source of various value-added products. In this regard, catalysts supported on carbon nanomaterials seem to outperform those supported on traditional supports. This was evident in the depolymerization of the components of lignocellulosic biomass such as cellulose, hemicellulose and lignin. Thus, under mild conditions, the yields of monomeric products obtained over catalysts supported on carbon nanosupports exceeded those obtained over other catalysts. Similarly, for the former catalysts, high activity and selectivity for the products of interest were also obtained during the conversion of vegetable oils and algae biomass components. Under more severe conditions, such as those encountered in HPR, the biomass-derived feeds are converted to hydrocarbons for transportation fuels. It should be noted that in all of these applications, the presence of water did not affect the performance of catalysts supported on carbon nanosupports.

Reactants containing α,β-conjugated C=C and C=O double bonds, such as cinnamaldehyde and citral, have been extensively investigated for the production of fine chemicals. In these applications, the high selectivity requirement for hydrogenation of the C=O bond while leaving the C=C bond intact could be achieved over noble metal catalysts supported on carbon nanosupports.

Hydrocarbons have been studied over catalysts supported on carbon nanomaterials either as individual reactants or as part of the feeds derived from conventional petroleum and also from nonconventional sources. The streams of alkenes (*e.g.* ethylene and styrene) from pyrolysis require a final polish to remove alkynes (*e.g.* acetylene and phenylacetylene) before further utilization. To ensure stability, the final traces of alkenes must be removed from gasoline and diesel fractions. Regulatory limits on the aromatics content of fuels can also be met by final polishing. In these applications, catalysts consisting of noble metals supported on CNTs and/or CNFs exhibited good performance at ambient temperature and with near atmospheric pressure H_2.

Results on the conversion of real feeds derived from petroleum and those obtained from nonconventional sources confirmed that with respect to activity and selectivity, catalysts supported on carbon nanomaterials exhibited better performance than those supported on conventional oxidic supports. Also, the performance of the former catalysts was better than that of the corresponding catalysts supported on amorphous carbons such as ACs and CBs. Therefore, great potential of carbon nanomaterials as supports for HYD and HPR catalysts is anticipated.

Edward Furimsky

List of Acronyms

AC	Activated carbon
AGO	Atmospheric gas oil
AR	Atmospheric residue
ASA	Amorphous silica–alumina
CALC	Cinnamyl alcohol
CALD	Cinnamaldehyde
CB	Carbon black
CNF	Carbon nanofiber
CNH	Carbon nanohorn
CNT	Carbon nanotube
CSTR	Continuous stirred tank reactor
CVD	Chemical vapor deposition
DBT	Dibenzothiophene
DMF	Dimethylfuran
EOR	End of run
FB	Fullerene black
FCC	Fluid catalytic cracking
FF	Furfural
FFA	Furfuryl alcohol
FLG	Few-layers graphene
FT	Fischer–Tropsch
GNP	Graphene nanoplatelet
GNR	Graphene nanoribbon
GO	Graphene oxide
GVL	γ-Valerolactone
HCALC	Hydrocinnamyl alcohol
HCALD	Hydrocinnamaldehyde
HCR	Hydrocracking

Catalysis Series No. 37
Carbon Nanomaterials in Hydrogenation Catalysis
By Edward Furimsky

Published by the Royal Society of Chemistry, www.rsc.org

HDAr	Hydrodearomatization
HDCl	Hydrodechlorination
HDM	Hydrodemetallization
HDN	Hydrodenitrogenation
HDO	Hydrodeoxygenation
HDP	Homogeneous deposition–precipitation
HDS	Hydrodesulfurization
HGO	Heavy gas oil
HiPCO	High-pressure carbon monoxide disproportionation
HIS	Hydroisomerization
HMF	Hydroxymethylfurfural
HPR	Hydroprocessing
HYD	Hydrogenation
id	Internal diameter
LA	Levulinic acid
LHSV	Liquid hourly space velocity
MF	Methylfuran
MWCNT	Multi-walled carbon nanotube
OCNT	Oxidized carbon nanotube
od	Outer diameter
rGO	Reduced graphene oxide
SWCNH	Single-walled carbon nanohorn
SWCNT	Single-walled carbon nanotube
TEM	Transmission electron microscopy
THF	Tetrahydrofuran
THFFA	Tetrahydrofurfuryl alcohol
TOF	Turnover frequency
TOS	Time on-stream
TPD	Temperature-programmed desorption
TPO	Temperature-programmed oxidation
VGO	Vacuum gas oil
VR	Vacuum residue
WABT	Weighted average bed temperature
WHSV	Weight hourly space velocity

Contents

Catalysis Series No. 37
Carbon Nanomaterials in Hydrogenation Catalysis
By Edward Furimsky

Published by the Royal Society of Chemistry, www.rsc.org

CHAPTER 1

Introduction

Carbon nanomaterials such as carbon nanotubes (CNTs), carbon nanofibers (CNFs), carbon nanohorns (CNHs) and fullerenes and graphene-derived solids such as graphene oxide (GO) and reduced graphene oxide (rGO) have been evaluated as potential supports for catalysts active during hydrogen transfer such as required for hydrogenation (HYD) and hydroprocessing (HPR) reactions.[1] HYD relates to the conditions employed during the HYD of various feeds and reactants for the production of fine chemicals and additives whereas HPR relates to the production of hydrocarbons such as those present in fuels and lubricants. There may be a significant difference between the operating conditions employed in these processes. For example, in HYD, the temperature rarely exceeds 150 °C, whereas in HPR, temperatures above 250 °C are necessary. Of course, owing to the complexity of feeds, an overlap of these temperature ranges may occur. Similar or the same catalysts have been used in these applications. Catalysts consisting of carbon nanosupports combined with noble metals (*e.g.* Pt, Pd, Ru, Rh and Ir), rare earth metals (*e.g.* Y, La and Ce) and transition metals (*e.g.* Ni, Co, Mo, W, Cu and Fe) have been the focus of attention.

Amorphous carbon materials such as carbon blacks (CBs) are included because the size of the particles may be in the nanorange. Activated carbons (ACs) are included whenever corresponding catalysts have been tested in comparison with catalysts supported on carbon nanomaterials. The applications of CBs and ACs in HPR catalysis, both as catalysts and as catalyst supports, were reviewed in detail more than a decade ago.[2] At that time, the database on similar applications of carbon nanomaterials was limited. Catalysts supported on oxidic supports are also included whenever they were tested in the same studies under similar conditions as those for catalysts supported on carbon nanosupports. This gives an opportunity to evaluate the rather unique performance of the latter catalysts in comparison with conventional catalysts.

Catalysis Series No. 37
Carbon Nanomaterials in Hydrogenation Catalysis
By Edward Furimsky

Published by the Royal Society of Chemistry, www.rsc.org

There are other potential applications of carbon nanomaterials. In catalysis, carbon nanomaterials have been evaluated both as catalysts and supports for catalysts in various non-HYD applications. These studies were reviewed by Serp and co-workers.[3,4] A detailed account of hydrogen-related applications of CNTs (*e.g.* production *via* electrolysis, storage) was provided by Orinakova and Orinak.[5] The potential of CNTs in several other energy-related applications was addressed in a review by Centi and Perathoner.[6] However these aspects, and many others, are outside the scope of the present review.

Feeds and reactants varying widely in structure and origin have been included in studies over catalysts supported on carbon nanomaterials. As the most abundant source of carbon, the components of lignocellulosic biomass such as cellulose, hemicellulose and lignin have attracted considerable attention as the feed for catalytic depolymerization and also a source of products requiring catalytic upgrading to either fine chemicals or fuels.[7] The conversions of cellulose and hemicellulose to glucose and fructose and to xylose, respectively, have been identified as important routes for the production of chemicals and fuels. Monomeric phenols and aromatics (*e.g.* guaiacol, eugenol and syringol) derived from lignin have been used as model compounds for HYD, HPR and hydrodeoxygenation (HDO).

Unique model compounds such as reactants containing α,β-conjugated C=C and C=O double bonds (*e.g.* cinnamaldehyde, citral and crotonaldehyde) have been studied. In these applications, catalysts supported on carbon nanosupports have been extensively investigated. This resulted from their selectivity for the HYD of C=O bonds while leaving C=C bonds intact. In this case, fine chemicals required for the production of pharmaceuticals, food additives, perfumery, *etc.*, were the products of interest. Conventional petroleum feeds and nonconventional feeds and also model compounds derived from these feeds have been used to evaluate the performance of catalysts supported on carbon nanosupports.

HYD is an important route for the conversion of carbon oxides (CO and CO_2) to useful products. Fischer–Tropsch (FT) synthesis is a method for the conversion of CO that has been used commercially for decades. In fact, the concept based on the pre-reduction of CO_2 to CO *via* the Boudart reaction using carbon or during the reduction with H_2:

$$CO_2 + C \rightarrow 2CO$$

$$CO_2 + H_2 \rightarrow CO + H_2O$$

followed by the utilization of CO as the feed for FT synthesis:

$$nCO + mH_2 \rightarrow \text{chemicals and fuels}$$

has merit because all steps involved are commercially proven. Studies in which either carbon nanomaterials alone or catalysts supported on carbon nanosupports, using CO and CO_2 as the feeds, have been noted. Because of its uniqueness, this concept deserves separate treatment to enhance the awareness of its merit.

In spite of the limited information available, the potential of catalysts supported on carbon nanomaterials in various environmentally related applications should be noted. For example, chlorinated solid waste (*e.g.* polychlorinated biphenyls) could be readily converted to hydrocarbons under fairly mild HYD conditions.[8] Nitrate contaminants could also be converted to benign products during HYD over these catalysts.[3,6] These few examples are introduced to indicate the suitability of catalysts supported on carbon nanomaterials for the conversion of hazardous wastes.

References

1. J. M. Planeix, N. Coustel, B. Coq, V. Brotons, P. S. Kumbhar, R. Dutartre, P. Geneste, P. Bernier and P. M. Ajayan, *J. Am. Chem. Soc.*, 1994, **116**, 7935.
2. E. Furimsky, *Carbons and Carbon Supported Catalysts in Hydroprocessing*, RSC Publishing, Cambridge, UK, 2008.
3. P. Serp, M. Corrias and P. Kalck, *Appl. Catal.*, 2003, **253**, 337–358.
4. P. Serp and J. S. Figueiredo, *Carbon Materials for Catalysis*, Wiley, Hoboken, New Jersey, 2009.
5. R. Orinakova and A. Orinak, *Fuel*, 2011, **90**, 3123–3140.
6. G. Centi and S. Perathoner, *ChemSusChem*, 2011, **4**, 913–925.
7. H. Li, A. Riisager, S. Saravanamurugan, A. Pandey, R. S. Sangwan, S. Yang and R. Luque, *ACS Catal.*, 2018, **8**, 148–187.
8. X. Guo, C. Yu, Z. Yin, S. Sun and C. T. Seto, *ChemSusChem*, 2018, **11**, 1617–1620.

CHAPTER 2

Properties of Carbons

Structurally, highly ordered nanomaterials in their pristine form, *i.e.* graphene, CNTs, CNFs, CNHs and fullerenes, and the corresponding modified materials are the primary objective of this book. Amorphous carbons, such as CBs and ACs, are briefly discussed because they have been included for comparison in many studies involving carbon nanomaterials. The focus is on the properties that are relevant for catalysis under HYD and HPR conditions and on those properties that can influence catalyst preparation, performance and stability.

2.1 Carbon Nanomaterials

Figure 2.1 shows graphite, which consists of a layered/planar structure. Only three layers are shown to indicate the covalent bonding between carbons (spheres) and the flow of van der Waals forces depicted by vertical lines. In reality, graphite consists of multiple layers of platelets. In this case, the two-dimensional layers are called graphene. It is evident that the layers are arranged as a honeycomb lattice. The distance between the planes is *ca.* 0.35 nm. Because the layers are held together by weak van der Waals forces only, they can be easily separated by various methods. One layer, *i.e.* graphene, is also shown in Figure 2.1. In this case, basal planes account for most of the surface area while the contribution of the edge regions to the total surface area is very small. In the diagram of graphene, the armchair and zigzag configurations of the edge regions are evident. A difference between the activities of basal planes and edges may be anticipated.[1] In the case of pristine graphene, the edges should anchor catalytically active metals more readily than basal planes. Also, the reactivity of the edges in contact with various reactants (*e.g.* HNO_3, H_2SO_4, H_2O_2) is much greater than that of the basal planes.

Catalysis Series No. 37
Carbon Nanomaterials in Hydrogenation Catalysis
By Edward Furimsky

Published by the Royal Society of Chemistry, www.rsc.org

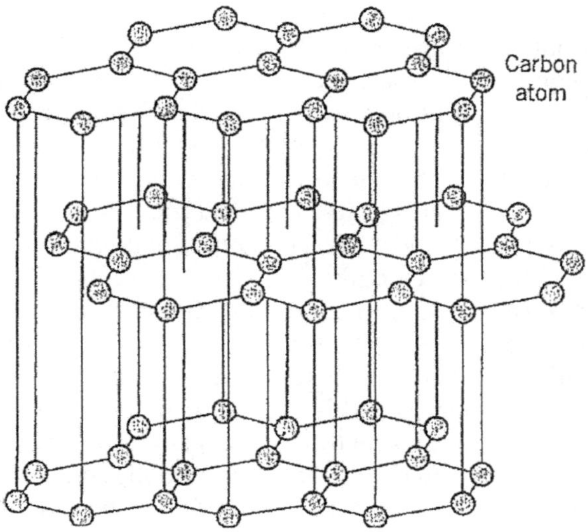

Graphite

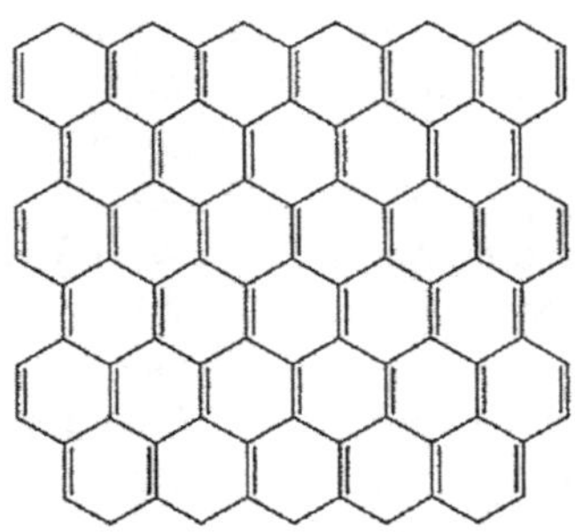

Graphene

Figure 2.1 Structures of graphite and pristine graphene.

It may be visualized that carbon nanosolids such as CNTs, CNFs and CNHs arose from the ability of the two-dimensional graphene sheets such as shown in Figure 2.1 to bend and/or to roll. This is supported by similar polycyclic, hexagonally arranged aromatic structures of basal planes in these nanomaterials. Similar methods have been used to increase the activity of basal planes in pristine carbon nanomaterials, *via* either functionalization or doping.

2.1.1 Graphene

As illustrated in Figure 2.1, pristine graphene is a single two-dimensional sheet of polycyclic, hexagonally arranged, sp^2-bonded aromatic carbon.[2–10] The nanosolid, referred to as few-layers graphene (FLG), consists of up to 10 stacked sheets of pristine graphene planes. It is usually a by-product of the production of monolayer graphene. The morphology of graphene is the reason for the large surface area. Thus, theoretically, the surface area of pristine graphene may approach 2500 $m^2 g^{-1}$.[6,7] The large surface area is favorable for strong adsorption of reactant molecules and also for anchoring catalytically active metals. However, because graphene is hydrophobic and nonreactive, functionalization of its surface is necessary for potential applications in catalysis.[8] During catalyst preparation, the anchoring of active metals is achieved much more readily using functionalized and/or doped graphene than with pristine graphene.

The most advanced method to obtain a higher yield of graphene is based on graphite oxide exfoliation.[11,12] Graphite oxide is prepared by the oxidation of graphite. The disruption of stacking in graphite oxide is necessary for graphene layer exfoliation. This can be achieved by ultrasonic, mechanical, chemical and thermal methods. The exfoliated oxidized layer is referred to as graphene oxide (GO). Recently, exfoliation *via* a rotor–stator method in supercritical CO_2 has been reported.[13] To be suitable as a catalyst support, reduction of GO to remove the excess of O-containing groups may be necessary. In this case, the product obtained is referred to as reduced graphene oxide (rGO). Figure 2.2 shows the tentative structures of GO and rGO. Various modifications of this method can be used and, in fact, have been adopted in some studies relevant for catalysis.[14] Graphene preparation using the chemical vapor deposition (CVD) method has also been applied. Details of this method applied below 1000 °C, with focus on the involvement of plasma and different carbon precursors and catalysts, were reported in a review by Naghdi *et al.*[15] Cossutta *et al.*[16] used a life-cycle approach in comparing these methods with the method based on the electrochemical exfoliation of graphite. Considering the industrial production of large quantities of graphene, graphite oxidation followed by reduction is the least impacting route when energy requirements and environmental issues are

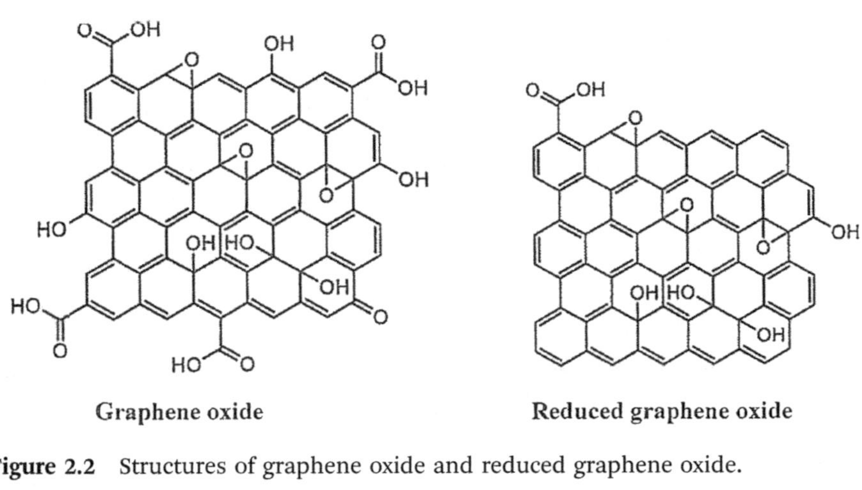

Figure 2.2 Structures of graphene oxide and reduced graphene oxide.

taken into consideration. The potential applications of graphene-supported catalysts for the HYD and HPR of various feeds have been extensively reviewed.[3,17,18]

2.1.2 Carbon Nanotubes (CNTs)

With respect to HYD and HPR catalysis, CNTs have been attracting more attention than any other carbon nanomaterial.[19–21] According to Su,[19] CNTs have been evaluated in catalysis both as supports and as catalysts. The rather unique selectivity at high conversions during industrially important reactions has been attributed to the efficient dispersions of active metals on the surface of CNTs. In addition, stable anchoring of metals on CNTs prevents catalyst deactivation due to sintering and losses of metals due to leaching during reaction. For some reactants, the curvature of the surface of basal planes in CNTs may influence interactions with active sites compared with the flat surface of graphene.

CNTs have a hexagonal geometry with sp^2 hybridization similar to that of graphite and/or graphene layers. Both single-walled carbon nanotubes (SWCNTs) and multi-walled carbon nanotubes (MWCNTs) have been the focus of attention.[22,23] The former can be visualized as a graphene sheet wrapped into a seamless cylinder with a wall thickness approaching the diameter of a carbon atom. The MWCNTs comprise several concentric SWCNTs, *i.e.* one concentric cylinder is located in another one with a larger diameter and so on. The spacing between the inner tubes in MWCNTs approaches that of the distance between graphene layers in graphite (*e.g.* ~0.34 nm).[24]

Figure 2.3 depicts SWCNTs and MWCNTs as visualized by Dodziuk and Dolgonos[25] and by Orinakova and Orinak,[24] respectively. The opened SWCNTs are terminated with edges, among which the armchair edge is metallic whereas zigzag and chiral edges are semiconductors. The surface

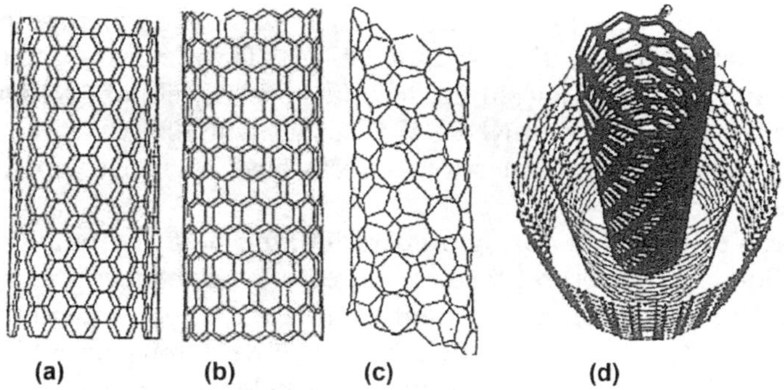

Figure 2.3 Structures of single-walled carbon nanotubes with (a) armchair, (b) zigzag and (c) chiral edges and (d) multi-walled carbon nanotube.

area of CNTs includes the basal planes of both internal and external surfaces and also the surface of edge regions. It is evident that the contribution of edges to the total surface area is rather small, although it can be increased by decreasing the length of the CNTs, *e.g.* on cutting in half the surface area of the edges is doubled. The internal basal planes can take part in catalysis provided that the inner diameter of the CNT is large enough to allow reactant molecules to enter and product molecules to exit. Similarly as in pristine graphene, the reactivity of basal planes in CNTs is much lower than that of the edge regions. For example, catalytically active metals would anchor with edges much more readily than with basal planes. However, similarly as for graphene, the reactivity of basal planes can be significantly increased by pretreatment such as functionalization and doping with heteroatoms.

Just a few examples of the methods for the preparation of CNTs are introduced. In this regard, a fairly extensive database can be accessed in the literature.[26–31] As examples, CVD and high-pressure carbon monoxide disproportionation (HiPCO) methods have been used for the preparation of CNTs.[26,27] The former is a batch process taking place in vacuum whereas HiPCO is a continuous process occurring in a flow of CO. In the presence of a catalyst, some control over the properties of CNTs in the HiPCO process can be maintained.[28,29] The ratio of the length to diameter of CNTs prepared by these methods can approach many millions to one. CNTs with a diameter as small as 1 nm can be prepared. In the CVD method described by Corrias *et al.*,[30] Fe(2.5%)/Al_2O_3 catalyst was fluidized in a fluidized bed reactor in a flow of $H_2 + N_2$. When a temperature of 650 °C had been attained, the flow rate of $H_2 + N_2$ was lowered and ethylene was introduced into the fluidized bed to attain an $N_2 : H_2$: ethylene volume ratio of about 1 : 2 : 1. Under these conditions, carbon yields exceeded 95% at a CNT selectivity approaching 100%. For the large-scale production of CNTs, modification of the CVD method such as in the spray pyrolysis of hydrocarbons (*e.g.* toluene) was recommended by Gomez *et al.*[31]

In several studies, waste plastics were identified as a potential source for the production of CNTs. In this case, in the first stage, a pyrolyzer is part of the multi-stage process developed by Yao *et al.*[32] The volatiles produced during pyrolysis entered a fixed bed of catalysts (*e.g.* Fe/α-Al_2O_3, Ni/γ-Al_2O_3, Ni/α-Al_2O_3 and Ni–Fe/γ-Al_2O_3) in the second stage. A yield of CNTs of almost 30 wt% was obtained over an Ni–Fe/Al_2O_3 catalyst at 800 °C. Other methods for the conversion of waste plastics to CNTs have been developed and described elsewhere.[33]

It has been evident that during preparation, the yield and quality of CNTs can be influenced by several factors, *i.e.* carbon source/precursor, type of catalyst and temperature. Subsequent pretreatment (*e.g.* functionalization and doping) of CNTs before the anchoring of active metals has an impact on the size and distribution of active metal particles and the stability of the catalyst. These aspects of CNTs were extensively discussed by Tessonnier and Su.[34]

2.1.3 Carbon Nanofibers (CNFs)

Cylindrical CNFs consist of graphene layers in the form of platelets and ribbons in which the layers are aligned perpendicular and parallel to the fiber axis, respectively. In addition, a spiral form of CNFs consists of platelets oriented parallel to the fiber axis with some bending along the axis. The visualization of these shapes of CNFs by Park and Baker is shown in Figure 2.4.[35] It can be seen that, in their structure, CNFs differ from CNTs, which are in the form of perfect cylinders. From a catalysis point of view, a significant contribution of the edge regions of graphene layers to the overall surface area of CNFs contrasts with the basal planes accounting for most of the surface area in CNTs (Figure 2.3). An increased spacing between graphene layers in the spiral CNFs should be noted. This may play a certain role during the anchoring of active metals compared with the platelet and ribbon forms of CNFs.

The size of CNFs varies widely, *i.e.* the diameter may be in the range 3–100 nm and the length varies from 0.1 to 1000 μm.[36] CNFs can be prepared by CVD in vacuum in the presence of a transition metal catalyst (most often Fe and Ni supported on γ-Al_2O_3). In this case, a source of carbon (*e.g.* methane, acetylene, ethylene) in a mixture with H_2 and He is decomposed at high temperature, *i.e.* ~1100 °C. It was reported that the morphology of CNFs can be influenced by the type of source of carbon and catalyst.[37,38] The carbon atoms generated that were deposited on the surface of the catalyst are the seeds for the growth of CNFs. The BET surface area of these CNFs may also vary. For example, the surface areas of the platelet, ribbon and spiral CNFs prepared by Park and Baker[35] shown in Figure 2.4 were 120, 85 and 45 $m^2\,g^{-1}$, respectively. The mechanistic aspects of the formation of CNFs on Ni-containing catalysts were investigated by Jarrah *et al.*[39] For manufacturing CNFs, both batch[36] and continuous process[40] have been developed. The features of these processes were described and compared by de Jong and Geus.[41]

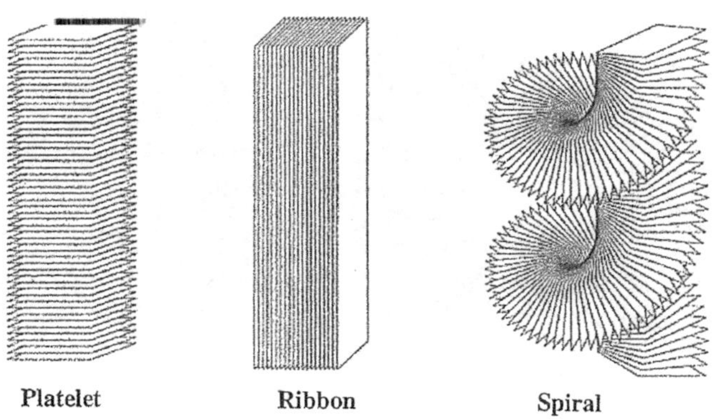

Figure 2.4 Structures of carbon nanofibers.

2.1.4 Carbon Nanohorns (CNHs)

With respect to applications as catalysts and catalyst supports, CNHs have attracted much less attention than the other carbon nanomaterials. CNHs consist of the both end-enclosed horn-shaped graphene sheets with a diameter of 2 nm at the tip and 2–4 nm at the wide area, for a single CNH with length varying from 30 to 50 nm. The CNHs have hexagonal geometry; however, the tip and bent bottom ends of CNHs may comprise pentagonal and heptagonal rings. This deviation from hexagonal coordination represents an irregularity that is responsible for higher reactivity at both ends of CNHs. Thus, during a controlled oxidation (*e.g.* in air at 693 K),[42] both ends of CNHs can be opened while the basal planes remain intact. Otherwise, if used as either catalysts or catalyst supports, reactants cannot diffuse in and reaction products out of the enclosed CNHs. In such a case, only the external surface of CNHs dominated by basal planes can be available. The CNH particles consist of the multi-CNH dahlia flower-type aggregates as shown in Figure 2.5 (left). The structure of the tip of CNHs is shown in Figure 2.5 (right). Experimentally, this morphology of CNHs aggregates can be imaged using transmission electron microscopy (TEM).

CNHs can be prepared by laser ablation of pure graphite and also by a simple pulsed arc discharge ignited between preheated carbon rods. Structurally, the product may contain several types of carbon. Thus, according to Utsumi *et al.*,[43] dahlia flower-type SWCNHs (Figure 2.5) prepared by CO_2 laser ablation of graphite under Ar comprised 2.5% of amorphous carbon, 15% of defective carbon at the cone part, 70% of tubular graphene carbon and 12% of graphitic carbon. A spherical bud-like shaped assembly is produced if the ablation is conducted under He. If present in the product, amorphous carbon can be readily removed by oxidation in air.[44,45] Controlled opening of CNHs could also be achieved by treatment with CO_2 but at higher temperatures than in air.[46,47]

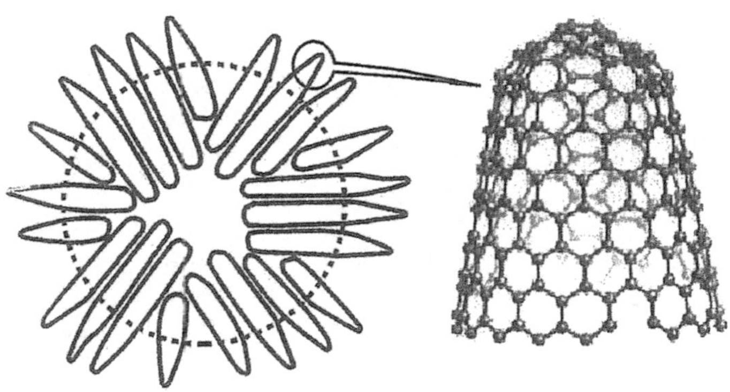

Figure 2.5 Structure of carbon nanohorn.

The porosity of CNH aggregates comprises the interstitial volume and the volume inside the single CNHs provided that the caps were opened by oxidation. Heat treatment at 693 K opens the intra-particle pores almost perfectly, leading to a micropore volume of 0.47 $mL\,g^{-1}$ and a specific surface area of 1010 $m^2\,g^{-1}$.[48] Subtraction of the N_2 adsorption isotherm of the SWCNH from that of the SWCNH treated at 693 K gave the N_2 adsorption isotherm only in the intra-particle pore spaces.[49] The adsorption sites derived from the difference in adsorption isotherms are assigned to pores having different interaction potentials. In pristine form, CNHs exhibit low reactivity, similar to that of pristine graphene, CNTs and CNFs. Therefore, pretreatment as applied to these nanomaterials is necessary to make CNHs suitable for application in catalysis.

2.1.5 Fullerenes

Fullerenes consist of a hollow sphere of carbons composed of hexagonal and pentagonal rings. The C_{60} form, shown in Figure 2.6, is the most common type of fullerenes. However, smaller (*e.g.* C_{20}, the smallest) and larger molecules (*e.g.* C_{70}, C_{72}, C_{76}, C_{84} and even C_{100}) also exist. Traces of fullerenes can be found in Nature, *e.g.* they can be formed in the atmosphere by lightning. Fullerenes can be produced by either an arc method or a combustion method. For the large-scale commercial production of fullerenes, the latter method is more suitable.[50] Fullerenes are present in the soot formed during incomplete combustion of carbonaceous materials. They can be isolated from the soot by extraction with an organic solvent. Fullerenes are insoluble in water but are readily soluble in common aromatic solvents (*e.g.* benzene, toluene, tetralin) and chlorinated aromatics.[51,52] Compared

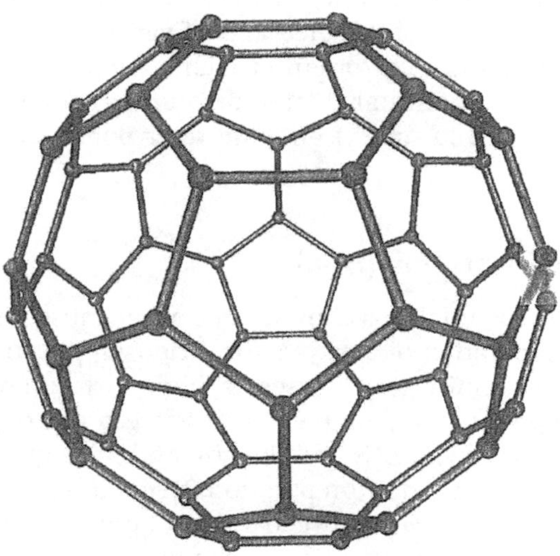

Figure 2.6 Structure of C_{60} fullerene.

with CNTs and CNFs, the information available on the performance of fullerenes as catalysts and/or catalyst supports under HYD conditions is much less extensive.

2.2 Amorphous Carbons

The potential of amorphous carbons such as CBs and ACs in HPR catalysis was extensively reviewed previously.[2] This included their role as catalysts and as supports for the catalysts. The performance of AC- and CB-supported catalysts was discussed in relation to catalysts supported on oxidic supports. In addition to low crystallinity, significantly different textural properties of ACs should be noted. For example, the surface area of ACs may be an order of magnitude greater than that of some carbon nanomaterials. Then, if used as supports, the effects of textural properties and those induced by the latter supports on active metals can be much more evident.

2.2.1 Carbon Blacks (CBs)

CBs are included for comparison with carbon nanomaterials, whenever appropriate. Thus, for several CBs, the particle size varies from 20 to 100 nm. In comparison with the highly ordered carbon nanomaterials discussed earlier, CB is an amorphous solid with a degenerate graphitic structure.[2] The irregularities of CBs are evident by the angular displacement of one layer with respect to another. The CB particles form aggregates with a concentric arrangement of layers. Chemically, the interior of the aggregates is less ordered than the exterior, hence it is more reactive. On exposure to air, all CBs chemisorb oxygen to form O-containing groups (*e.g.* peroxy, carboxylic, quinonic, phenolic). Commercially, CBs are manufactured by thermal decomposition of methane (natural gas) and of various petroleum fractions. Depending on the process (*e.g.* origin of hydrocarbon feed, features of the reactor, catalytic *versus* noncatalytic, temperature), the surface area of CBs can vary from 100 to 700 $m^2 g^{-1}$ and the total pore volume from 0.8 to 1.7 $mL g^{-1}$.[2]

2.2.2 Activated Carbons (ACs)

Among all carbon materials, ACs have been used most frequently as the support for the preparation of catalysts for various applications. Because of the rather low crystallinity, ACs possess a high surface area with a large number of micropores. For example, the surface area frequently exceeds 1000 $m^2 g^{-1}$. The porosity of ACs and the size and form of the particles can be influenced by the method of preparation. Depending on the conditions, the pore size of an AC may vary between 1 and 10 nm. However, pores larger than 10 nm can be also present. Such macropores serve as channels for reactants diffusing into micropores.

ACs are prepared by the pyrolysis of organic matter (*e.g.* petroleum coke, coal, biomass, refinery residues) at about 900 °C in the first stage. After driving off the volatiles, the solid residue is activated in the second stage in the presence of diluted air, steam and/or CO_2. This suggests that the surface of the AC is oxidized, in contrast to that of pristine carbon nanomaterials. However, because of the temperatures employed during the preparation, the concentration of surviving O-containing groups on the surface of the AC may be limited. Therefore, functionalization of the AC at lower temperatures may be beneficial before it can be used for catalyst preparation.

ACs are manufactured in the form of powder, granules, pellets and extrudates. This suggests that there is a choice of suitable shapes and sizes of AC particles for catalyst preparation for HYD and HPR applications.[2] Other fields of catalysis in which ACs have been used as a support are outside the scope of this review. Of course, there are numerous noncatalytic applications of ACs, particularly in the area of safety and the environment.

References

1. L. Rodríguez-Pérez, M. Ángeles Herranza and N. Martín, *Chem. Commun.*, 2013, **49**, 3721–3735.
2. N. G. Sahoo, Y. Z. Pan, L. Li and S. H. Chan, *Adv. Mater.*, 2012, **24**, 4203–4210.
3. E. Furimsky, *Ind. Eng. Chem. Res.*, 2017, **56**, 11359–11371.
4. A. K. Geim, *Science*, 2009, **324**, 1530–1534.
5. A. K. Geim and K. S. Novoselov, *Nat. Mater.*, 2007, **6**, 183–191.
6. A. H. C. Neto, F. Guinea, N. M. R. Peres, K. S. Novoselov and A. K. Geim, *Rev. Mod. Phys.*, 2009, **81**, 109–162.
7. C. Soldano, A. Mahmood and E. Dujardin, *Carbon*, 2010, **48**, 2127–2150.
8. D. R. Haag and H. H. Kung, *Top. Catal.*, 2014, **57**, 762–773.
9. D. C. Elias, R. R. Nair, T. M. G. Mohiuddin, S. V. Morozov, P. Blake, M. P. Halsall, A. C. Ferrari, D. W. Boukhvalov, M. I. Katsnelson, A. K. Geim and K. S. Novoselov, *Science*, 2009, **323**, 610–613.
10. L. Li, C. Liang, Z. Feng, P. Ying, D. Wang and C. Li, *J. Mol. Catal. A: Chem.*, 2004, **211**, 103–109.
11. W. S. Hummers and R. E. Offeman, *J. Am. Chem. Soc.*, 1958, **80**, 1339–1342.
12. M. P. Lavin-Lopez, J. L. Valverde, L. Sanchez-Silva and A. Romero, *Ind. Eng. Chem. Res.*, 2016, **55**, 845–855.
13. Y. Gai, W. Wang, D. Xiao, H. Tan, M. Lin and Y. Zhao, *Ind. Eng. Chem. Res.*, 2018, **57**, 8220–8229.
14. C. Chen, Q. H. Yang, Y. Yang, W. Lv, Y. Wen, P. X. Hou, M. Wang and H. M. Cheng, *Adv. Mater.*, 2009, **21**, 3007–3011.
15. S. Naghdi, K. Y. Rhee and S. J. Park, *Carbon*, 2018, **127**, 1–12.
16. M. Cossutta, J. McKechnie and S. J. Pickering, *Green Chem.*, 2017, **19**, 5874.

17. M. V. Klyuev, N. A. Magdalinova and P. A. Kalmykov, *Pet. Chem.*, 2016, **56**, 1093–1106.
18. B. C. Thompson, E. Murray and G. G. Wallace, *Adv. Mater.*, 2015, **27**, 7563–7582.
19. D. S. Su, *ChemSusChem*, 2011, **4**, 811–813.
20. S. Iijima and T. Ichihashi, *Nature*, 1993, **363**, 603.
21. S. Iijima, *Nature*, 1991, **354**, 56–58.
22. K. Takeuchi, T. Hayashi, Y. A. Kim, K. Fujisawa and M. Endo, *Nanosyst.: Phys., Chem., Math.*, 2014, **5**, 15–24.
23. G. Stando, D. Łukawski, F. Lisiecki, Filip and D. Janas, *Appl. Surf. Sci.*, 2019, **463**, 227–233.
24. R. Orinakova and A. Orinak, *Fuel*, 2011, **90**, 3123–3140.
25. H. Dodziuk and G. Dolgonos, *Chem. Phys. Lett.*, 2002, **356**, 79–83.
26. N. Karousis, N. Tagmatarchis and D. Tasis, *Chem. Rev.*, 2010, **110**, 5366–5397.
27. A. Eatemadi, H. Daraee, H. Karimkhanloo, M. Kouhi, N. Zarghami, A. Akbarzadeh, M. Abasi, Y. Hanifehpour and S. W. Joo, *Nanoscale Res. Lett.*, 2014, **9**, 393.
28. M. J. Bronikowski, P. A. Willis, D. T. Colbert, K. A. Smith and R. E. Smalley, *J. Vac. Sci. Technol., A*, 2001, **19**, 1800–1805.
29. P. Nikolaev, *J. Nanosci. Nanotechnol.*, 2004, **4**, 307–316.
30. M. Corrias, B. Caussat, A. Ayral, J. Durand, Y. Kihn, P. Kalck and P. Serp, *Chem. Eng. Sci.*, 2003, **58**, 4475–4482.
31. J. Gomez, Y. Verde, J. Lara-Romero and G. Alonso-Nuñez, *Fullerenes, Nanotubes, Carbon Nanostruct.*, 2009, **17**, 507–518.
32. D. Yao, Y. Zhang, P. T. Williams, H. Yang and H. Chen, *Appl. Catal., B*, 2018, **221**, 584–597.
33. A. Bazargan and G. McKay, *Chem. Eng. J.*, 2012, **195/196**, 377–391.
34. J.-P. Tessonnier and D. S. Su, *ChemSusChem*, 2011, **4**, 824–847.
35. C. Park and R. T. K. Baker, *J. Phys. Chem. B*, 1998, **102**, 5168–5177.
36. G. Tibbetts, M. Lake, K. Strong and B. Rice, *Compos. Sci. Technol.*, 2007, **67**, 1709–1718.
37. T.-J. Zhao, D. Chen, Y.-C. Dai, W.-K. Yuan and A. Holmen, *Top. Catal.*, 2007, **45**, 87.
38. N. M. Rodriguez, A. Chambers and R. T. K. Baker, *Langmuir*, 1995, **11**, 3862–3866.
39. N. B. Jarrah, J. G. van Ommen and L. Lefferts, *J. Catal.*, 2006, **239**, 460–469.
40. P. J. F. Harris, *Carbon Nanotube Science*, Cambridge University Press, Cambridge, UK, 2009.
41. K. P. de Jong and J. W. Geus, *Catal. Rev.: Sci. Eng.*, 2000, **42**, 481–510.
42. E. Bekyarova, K. Kaneko, D. Kasuya, K. Murata, M. Yudasaka and S. Iijima, *Langmuir*, 2002, **18**, 4138–4141.
43. S. Utsumi, J. Miyawaki, H. Tanaka, Y. Hattori, T. Itoi, N. Ichikuni, H. Kanoh, M. Yudasaka, S. Iijima and K. Kaneko, *J. Phys. Chem. B*, 2005, **109**, 14319–14324.

44. S. Iijima, M. Yudasaka, R. Yamada, S. Bandow, K. Suenaga, F. Kokai and K. Takahashi, *Chem. Phys. Lett.*, 1999, **309**, 165–170.
45. D. Kasuya, M. Yudasaka, K. Takahashi, F. Kokai and S. Iijima, *Proc. the 21th Fullerene Symp. Tsukuba, Japan*, 2001, p. 27.
46. E. Bekyarova, K. Kaneko, M. Yudasaka, D. Kasuya, S. Iijima, A. Huidobro and F. Rodriguez-Reinoso, *J. Phys. Chem. B*, 2003, **107**, 4479–4484.
47. J. Fan, M. Yudasaka, J. Miyawaki, K. Ajima, K. Murata and S. Iijima, *J. Phys. Chem. B*, 2006, **110**, 1587–1591.
48. K. Murata, K. Kaneko, W. Steele, F. Kokai, K. Takahashi, D. Kasuya, K. Hirahara, M. Yudasaka and S. Ijima, *J. Phys. Chem. B*, 2001, **105**, 10210–10216.
49. T. Yamaguchi, S. Bandow and S. Iijima, *Chem. Phys. Lett.*, 2004, **389**, 181–185.
50. M. A. Wilson, L. S. K. Pang, G. D. Willett, K. J. Fisher and I. G. Dance, *Carbon*, 1992, **30**, 675–693.
51. K. N. Semenov, N. A. Charykov, V. A. Keskinov, A. K. Piartman, A. A. Blokhin and A. A. Kopyrin, *J. Chem. Eng. Data*, 2010, **55**, 13–36.
52. A. V. Talyzin and I. Engström, *J. Phys. Chem. B*, 1998, **102**, 6477–6481.

CHAPTER 3

Modifications of Carbon Nanomaterials

As indicated earlier, the highly ordered graphitic structure of pristine carbon nanomaterials (*e.g.* graphene, CNTs, CNFs, CNHs and fullerenes) is the reason for their high hydrophobicity. In the pristine form, basal planes, accounting for most of the surface area of these solids, are unreactive. Only edge planes, accounting for a small fraction of total surface area, may possess some reactivity. To find applications in catalysis, the reactivity of carbon nanomaterials must be enhanced significantly. This can be achieved either by replacing some C atoms in the graphitic network with other atoms (*e.g.* S, N, O, P, B) or introducing functional groups (*e.g.* amino, hydroxyl, carboxylic, sulfonic); these approaches are termed doping and functionalization, respectively. By applying these methods to carbon nanomaterials, the reactivity and slurryability in water and various solvents are significantly enhanced.

Modifications of the surface chemistry of carbon nanomaterials by functionalization and doping were discussed extensively by Bandosz.[1] In the following, the emphasis is on those aspects of modifications that play a crucial role during the preparation of catalysts. First, modifications of the pristine carbons are essential for anchoring active metals on the surface of these solids and the subsequent catalyst performance during utilization. Thus, after modification, the hydrophilicity of carbons is increased and the catalyst preparation (*e.g.* aqueous-phase impregnation) is simplified. Of particular importance is the significantly enhanced stability of the catalysts supported on modified carbon nanosupports compared with those supported on pristine carbon nanomaterials. This has been confirmed by the negligible decline in catalyst activity and leachability of active metals during repeated catalytic cycles.

Catalysis Series No. 37
Carbon Nanomaterials in Hydrogenation Catalysis
By Edward Furimsky

Published by the Royal Society of Chemistry, www.rsc.org

3.1 Functionalization

Acidic groups varying in acid strength (*e.g.* –OH, –COOH, –SO_3H) and basic groups (*e.g.* –NH_2) can be introduced into the structure of carbon nanomaterials. Subsequently, the reactivity and hydrophilicity are significantly enhanced compared with pristine carbons. After the first stage of preparation, purification may be necessary to obtain high-quality carbon nanomaterials. The objective of purification is to remove amorphous carbon carryovers and mineral matter such as the remains of the catalyst used during preparation. Detailed accounts of the methods of purification, *i.e.* chemical, physical and their combinations, were given by Hou *et al.*[2] It was evident that some O-containing groups could be introduced already during the purification stage.

Both liquid- and gas-phase methods have been used for functionalization *via* the oxidation of carbon nanomaterials. In liquid-phase oxidation, HNO_3 has been the most frequently used oxidant. In this case, both the oxidation and purification can be achieved simultaneously by refluxing with a solution of HNO_3. Other oxidants for liquid-phase oxidation include H_2O_2, HCl, H_2SO_4, HNO_3 and $KMnO_4$.[2] Gas-phase oxidants include air, steam, Cl_2 and CO_2. After oxidation, the concentration of O-containing groups on the surface can be optimized by subsequent reduction, which can be readily achieved either thermally or in a flow of gases such as H_2 and/or in an inert gas.

In the case of graphene supports, a high degree of functionalization can already be achieved during the preparation stage, provided that graphite oxide was the starting material for the exfoliation method employed.[3–6] In fact, the layer of graphene oxide (GO) obtained may be over-oxidized. A suitable support can be obtained by the controlled reduction of GO to obtain a reduced GO (rGO). The difference between the concentrations of O-containing groups on GO and rGO is illustrated in Figure 2.2.

During oxidation by treatment with HNO_3, O-containing groups (*e.g.* hydroxyl, carboxyl, ether) are introduced. Hauke and Hirsch[7] showed that fairly complex O-containing groups can be formed on the surface of carbon nanomaterials. At temperatures approaching 900 °C, a desirable level of oxidation can also be achieved using either steam or CO_2 as oxidant. The potential of Cl_2 as a gas-phase oxidant has not been fully explored. Most frequently, the pretreatment of carbon nanomaterials has been carried out using concentrated HNO_3 (68%) at 110 °C under reflux. Deng and co-workers[8,9] investigated the effect of varying HNO_3 concentration on the surface acidity of CNTs as determined by NH_3 adsorption. Figure 3.1 confirms the increase in the amount of NH_3 adsorbed with increasing concentration of HNO_3. This corresponds to the increasing concentration of carboxylic groups on the CNT surface. The desorption of NH_3 at ~210 °C was attributed to weak acidic sites and the broad peak at 450–600 °C to medium-strength acidic sites.

The caps present in CNHs, and also maybe present in CNTs, can be opened by controlled oxidation under mild conditions. Consequently, the exposed

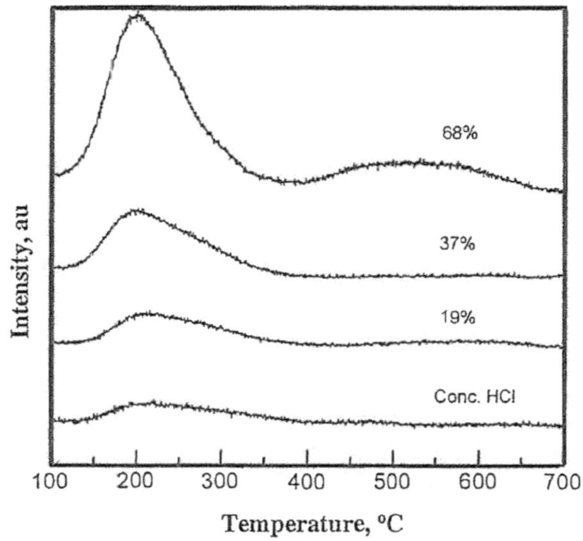

Figure 3.1 Effect of nitric acid concentration on the NH_3 TPD profile. Reproduced from ref. 7 with permission from Springer Nature, Copyright 2009.

end carbons are much more reactive than the carbons in the walls represented by basal planes. Therefore, the functionalization in the edge regions of CNTs, CNFs and CNHs and also that of peripheral carbons in graphene are much more extensive than that in the walls represented by basal planes. Similarly, the exposed edges of all types of CNFs such as shown in Figure 2.4 can be oxidized much more readily than the corresponding basal planes. These issues were discussed extensively by Moradi *et al.*[10] The functionalization of CNFs using either HNO_3 or HNO_3–H_2SO_4 resulted in a change in textural properties while the chemical structure remained unaltered.[11] Presumably, the removal of amorphous carbon and remaining mineral matter from CNFs resulted in increases in surface area and pore volume. These improvements in textural properties became more evident with increasing strength of the acid solution. For CNTs and CNHs, the improvement in textural properties was associated with the opening of end caps.

In a study by Ribeiro *et al.*,[12] the functionalization and/or purification of CNTs were carried out in a Soxhlet apparatus by refluxing with 6 M nitric acid for 3 h to obtain sample CNT1. Another oxidation was carried out in a round-bottomed flask using the same concentration acid solution to obtain sample CNT2, which was then thermally treated in a flow of N_2 at 400 °C (CNT2-400), 700 °C (CNT2-700) and 900 °C (CNT2-900). Table 3.1[12] shows that oxidation in the flask was much more efficient than that in a Soxhlet apparatus. Thermal treatment had the most pronounced effect on the removal of carboxylic and anhydride groups whereas carbonyl groups were least affected.

Table 3.1 Concentration of O-containing groups on functionalized CNTs.[12]

	O-containing group/$\mu mol\,g^{-1}$				
Sample	Carboxylic	Anhydride	Lactone	Phenol	Carbonyl
CNT1	105	71	57	399	380
CNT2	373	247	116	879	907
CNT2-400	33	149	71	508	617
CNT2-700	—	—	29	122	326

Table 3.2 TPD distribution of O-containing groups of nitric acid-treated MWCNTs.[13]

Functional groups	CO/$\mu mol\,g^{-1}$	CO_2/$\mu mol\,g^{-1}$
MWCNT–HNO_3	2130	1300
1. Strong carboxylic acids (560 K)	129	580
2. Weak carboxylic acids (700 K)	—	440
3. Anhydrides (790 K)	160	160
4. Lactones (935 K)	—	110
5. Phenols (885 K)	920	—
6. Carbonyl/quinone (1050 K)	920	—
MWCNT–original	210	60

Machado *et al.*[13] gave an excellent account of the surface modifications incurred by CNTs on treatment with nitric acid. CNTs were refluxed with 7 M nitric acid and the significant increase in the oxygen content of the CNTs that occurred during the treatment was associated with the formation of O-containing groups. This was clearly confirmed by the large amounts of CO_2 and CO released during the temperature-programmed desorption (TPD) of the nitric acid-treated CNTs compared with a negligible amount released with the untreated CNTs. The following O-containing groups were identified as strong (1) and weak (2) carboxylic acids, anhydrides (3), lactones (4), phenols (5) and ethers and carbonyls/quinones (6). Therefore, acidity can be introduced to the originally inert surface of CNTs by nitric acid treatment. An attempt to quantify these groups gave the results shown in Table 3.2.[13] The role of O-containing groups introduced on CNFs was described by Plomp *et al.*[14] and theoretical and practical aspects of the functionalization of CNFs were discussed by Rokhina *et al.*[15]

According to Pagona *et al.*,[16] carboxylic groups created on a carbon nanomaterial surface can be readily converted to acyl chloride, *e.g.* by reaction with $SOCl_2$. The acyl chloride can be further converted to amides, esters and thioesters *via* reaction with an amine, alcohol and thioalcohol, respectively. Therefore, in addition to typical O-containing groups, other functional groups can be introduced in carbon nanomaterials.

The presence of Brønsted acidity is necessary for carbon nanomaterials to be considered as supports for bifunctional catalysts. In addition to hydroxyl and carboxylic groups, stronger Brønsted acidity can be introduced by sulfonation. Details of the methods used to introduce sulfonyl groups in carbon nanomaterials were reported by Oger *et al.*[17] and Chen *et al.*[18] Then, under suitable conditions (*e.g.* presence of water), a proton can be generated by the dissociation of sulfonyl groups, *i.e.*

$$\text{-SO}_3\text{H} \rightarrow \text{-SO}_3^- + \text{H}^+$$

Although less efficiently, protons can also be donated by carboxylic and phenolic groups if present on the surface of carbon nanomaterials. Therefore, originally pristine nonreactive carbon nanomaterials can be functionalized to become efficient sources of Brønsted acidity. A composite consisting of CNFs and a sulfated perfluoropolymer (*e.g.* Nafion) exhibited strong surface acidity due to the presence of very stable sulfonic groups. The preparation of such a composite was described by Koskina *et al.*[19]

Experimental procedures to functionalize CNTs and CNFs by sulfonation were described by Liu *et al.*[20] In this case, carbon nanomaterials were grafted with either *p*-styrenesulfonic acid or benzenesulfonic acid. For preparation, sodium *p*-styrenesulfonate was dissolved in deionized water and sonicated with either CNTs or CNFs for 1 h. The mixture was stirred with potassium persulfate, added as a radical initiator. The solid obtained was filtered, washed and refluxed with concentrated HCl. For the preparation of benzenesulfonic acid-grafted solids, a mixture of CNTs and 4-benzenediazonium sulfonate in distilled water and ethanol was cooled to 5 °C followed by the addition of H_3PO_2. After stirring at 5 °C, the solid was filtered off, washed with water followed by acetone and oven dried at 80 °C. The structures of CNTs sulfonated by these two sulfonating agents are depicted in Figure 3.2.[20] Zhang *et al.*[21] prepared perfluorosulfonic acid-functionalized CNTs using the deposition of liquid perfluorosulfonic acid–polytetrafluoroethylene copolymer on CNTs. Such CNTs exhibited high stability in both polar and nonpolar solvents.

An interesting method for modifying the surface of CNFs and CNTs was described by Castillejos *et al.*[22] The objective was to graft long alkyl chains on the surface of carbon nanomaterials. First, the CNFs and CNTs were treated with nitric acid to introduce carboxylic groups to the surface. The functionalized solids were then chlorinated by refluxing in a solution of

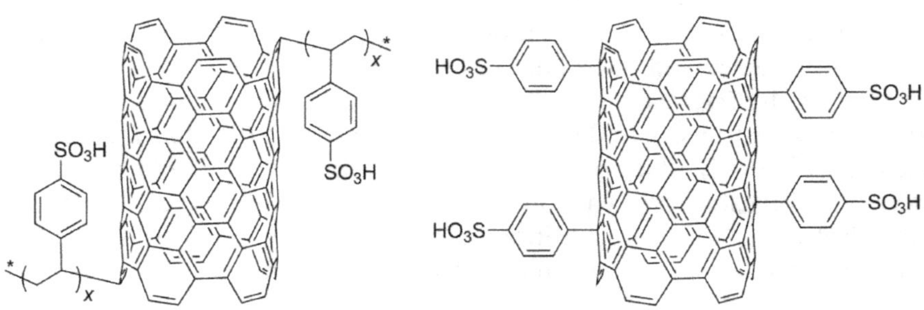

Figure 3.2 Structures of sulfonated carbon nanotubes.
Reproduced from ref. 20 with permission from the Royal Society of Chemistry.

thionyl chloride at 343 K for 24 h. After evaporating the solvent and vacuum drying, the chlorinated CNFs and CNTs were reacted with hexadecylamine in THF at 338 K for 24 h. This resulted in the formation of an amide surface group with a long alkyl chain attached, *i.e.*

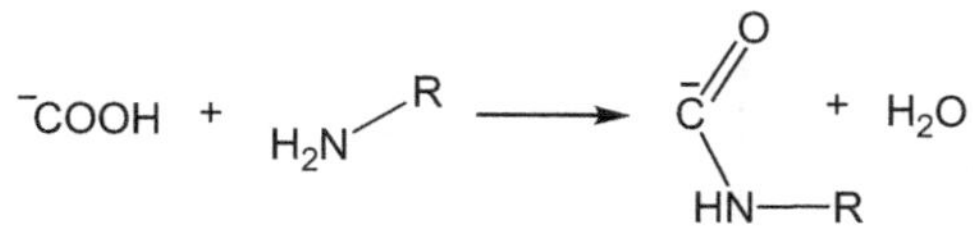

The amide entity is attached to the surface of CNFs and CNTs. In parallel, the grafting was performed with propylamine to obtain a propylamide group.

Li *et al.*[23] described a method for the preparation of multifunctional CNTs. In this case, an $-SO_3H$ group was introduced first followed by an $-NH_2$ group. The bifunctional SO_3H-NH_2 CNTs could be further functionalized with a chromium species to enhance the acidity of the support. This suggests that stable CNTs with coexisting basic and acidic groups can be prepared. The potential of such supports in the preparation of bifunctional catalysts may be anticipated.

The functionalization of CNHs can be performed in similar manner to that of other carbon nanomaterials. Thus, on oxidative treatment, the cone end of CNHs can be readily opened. The open ends of CNHs can be functionalized by nitric acid to introduce carboxylic groups.[24,25] These groups can be converted to acyl chlorides (CNH-COCl) by treatment with thionyl chloride and/or oxalyl chloride. The reactivity of acyl chlorides attached to CNHs may give rise to several other entities, *e.g.* amides, esters and thioesters. This suggests that the solubility of CNHs can be modified by functionalization to suit different solvents. In another method, CNHs were immersed in 95% sulfuric acid for 24 h.[26] Subsequently, the CNHs were washed with distilled water, oven dried at 110 °C for 12 h and calcined at 400 °C for 3 h.

3.2 Doping

Replacing carbon atoms with other atoms, termed doping, can disrupt the highly ordered hexagonal structure of pristine carbon nanomaterials. Because dopants (*e.g.* N, S, B, P) have a different electronegativity to carbon, the electronic properties of the surface are changed on doping.[27] Consequently, if used alone, the reactivity and catalytic activity of the doped carbon nanomaterials are increased. Moreover, the anchoring of active metals on the doped carbon nanosupports can be achieved more readily compared with that on the undoped supports. Apparently, the methodology for the doping of graphene, CNTs, CNFs and CNHs is similar.[28]

In a study by Chen *et al.*,[29] the doping of GO with nitrogen was achieved by thermal treatment of GO with urea. First, GO was dispersed in water by sonication before the addition of urea, then the mixture was stirred continuously until the water had evaporated. The solid was then calcined at various temperatures. In this case, N-containing structures such as graphitic, pyridinic and pyrrolic were formed. For example, after thermal treatment at 600 °C, the resulting GO contained 16.0 wt% of nitrogen. In this case, pyridinic, pyrrolic and graphitic nitrogen accounted for 38.4, 31.8 and 25.9%, respectively. Phenanthroline is another agent that can be used as the source of nitrogen for the doping of carbon nanosupports.[30]

Guellati *et al.*[31] described a catalytic CVD method based on the simultaneous decomposition of ethane and ammonia (at 750 and 850 °C) over an Fe/Al_2O_3 catalyst. The method produced MWCNTs consisting of concentric shells with external walls doped with nitrogen. In one mode, MWCNTs were formed in a flow of ethane and ammonia, while the other mode consisted of the formation of MWCNTs first without ammonia followed by the presence of both ethane and ammonia. The latter mode produced CNTs with nitrogen strongly bound with carbon as part of the pyridinic and quaternary coordination in contrast to the CNTs produced during the decomposition of both ethane and ammonia being present during the entire run.

The preparations of single-, double- and triple-doped CNTs were described by Liu *et al.*[32] Nitrogen-doped CNTs (N-CNTs) can be prepared by mixing 3,4-dihydroxyphenethylamine, referred to as dopamine (DA), with CNTs in buffer (pH 8.5). The polymerization of DA on the surface of CNTs gave polyDA (PDA). This was converted to multi-walled N-CNTs by pyrolysis under N_2. Double- and triple-doped CNTs can be obtained by depositing different precursors on the reactive PDA layer on the surface of CNTs. For example, 2-mercaptoethanol (2-ME) and 3-aminobenzeneboronic acid (3-ABBA) were used as the source of S and B to prepare S–N-CNTs and B–N-CNTs, respectively. As the source of both S and B, 4-mercaptophenylboronic acid was used to obtain triple-doped N–S–B-CNT catalysts (N–S–B-CNTs) using the same synthesis strategy. Elemental analyses and textural properties of the doped CNTs are given in Table 3.3.[32] The potential of doping and/or replacing carbon atoms in fullerenes and CNTs with N and B was explored by Tenne.[33]

Table 3.3 Composition and textural properties of single-, double- and triple-doped CNTs.[32]

	CNTs	N-CNTs	N–S-CNTs	N–B-CNTs	N–S–B-CNTs
Analysis/wt%					
C		98.8	99.1	93.0	91.5
N		1.3	0.6	6.0	2.3
S		—	0.3	—	0.3
B		—	—	1.0	5.9
Textural					
$SA^a/m^2\,g^{-1}$	130	126	130	122	82
$PV^a/cm^3\,g^{-1}$	0.51	0.44	0.6	0.51	0.41
PS^a/nm	15.7	19.7	22.5	17.0	23.7

[a]SA, surface area; PV, pore volume; PS, pore size.

In a study by Marco *et al.*,[34] CNFs were doped with B, P and N. In this case, B-doped CNFs with various amounts of B were prepared *via* incipient wetness impregnation using an aqueous solution of boric acid. Then, the B-CNFs were calcined at 773 K in N_2. P-CNFs were prepared in a similar way by impregnation with ammonium phosphate and triphenylphosphine. For doping with N, the CNFs were contacted with a flow of ethylenediamine diluted with Ar at 1023 K.

Louisia *et al.*[35] described a novel method for the doping of CNTs with sulfur. In this case, the growth of CNTs on an Fe catalyst was initiated in a flow of ethylene as carbon source. Subsequently, ethylene was replaced with thiophene as the source of sulfur. Up to 7 wt% of sulfur could be introduced into CNTs using this method.

In summary, the properties of carbon nanosupports can be significantly modified by either doping or functionalization. There is little information on applying both treatments on the same carbon support. For example, functionalization of the previously doped solids may further modify their properties, although it adds to their complexity. Nevertheless, compared with the originally inactive pristine carbon nanomaterials, the activity increased significantly after either functionalization or doping.[36–41] This was confirmed by increased adsorption of various reactants and also increased catalytic activity under HYD conditions. Details of these observations are given in the following chapters.

References

1. T. J. Bandosz, Surface Chemistry of Carbon Materials, in *Carbon Materials for Catalysis*, ed. P. Serp and J. S. Figueiredo, Wiley, Hoboken, New Jersey, 2009.
2. P.-X. Hou, C. Liu and H.-M. Cheng, *Carbon*, 2008, **46**, 2003–2025.
3. E. Furimsky, *Ind. Eng. Chem. Res.*, 2017, **56**, 11359–11371.
4. W. S. Hummers and R. E. Offeman, *J. Am. Chem. Soc.*, 1958, **80**, 1339–1342.
5. M. P. Lavin-Lopez, J. L. Valverde, L. Sanchez-Silva and A. Romero, *Ind. Eng. Chem. Res.*, 2016, **55**, 845–855.
6. C. Chen, Q. H. Yang, Y. Yang, W. Lv, Y. Wen, P. X. Hou, M. Wang and H. M. Cheng, *Adv. Mater.*, 2009, **21**, 3007–3011.
7. F. Hauke and A. Hirsch, Covalent Functionalization of Carbon Nanotubes, in *Carbon Nanotubes and Related Structures: Synthesis, Characterization, Functionalization and Applications*, Wiley, 2010, ch. 6, p. 135.
8. W. Deng, X. Tan, W. Fang, Q. Zhang and Y. Wang, *Catal. Lett.*, 2009, **133**, 167–174.
9. W. Deng, Y. Wang, Q. Zhang and Y. Wang, *Catal. Surv. Asia*, 2012, **16**, 91–105.
10. O. Moradi, M. Yari, K. Zare, B. Mirza and F. Najafi, *Fullerenes, Nanotubes, Carbon Nanostruct.*, 2012, **20**, 138–151.

11. M. L. Toebes, J. M. P. van Heeswijk, J. H. Bitter, A. J. Van Dillen and K. P. De Jong, *Carbon*, 2004, **42**, 307–315.
12. L. S. Ribeiro, J. J. Delgado, J. J. de Melo Órfão and M. F. Ribeiro Pereira, *ChemCatChem*, 2017, **9**, 888–896.
13. B. F. Machado, H. T. Gomes, P. Serp, P. Kalck and J. L. Faria, *ChemCatChem*, 2010, **2**, 190–197.
14. A. J. Plomp, H. Vuori, A. O. I. Krause, K. P. de Jong and J. H. Bitter, *Appl. Catal.*, 2008, **351**, 9–15.
15. E. V. Rokhina, M. Lahtinen, K. Makarova, V. Jegatheesan and J. Virkutyte, *Bioresour. Technol.*, 2012, **113**, 127–131.
16. G. Pagona, N. Tagmatarchis, J. Fan, M. Yudasaka and S. Iijima, *Chem. Mater.*, 2006, **18**, 3918–3920.
17. N. Oger, Y. F. Lin, C. Labrugère, E. Le Grognec, F. Rataboul and F.-X. Felpin, *Carbon*, 2016, **96**, 342–350.
18. J. Cheng, Y. Qiu, J. Zhang, R. Huang, W. Yang and Z. Fan, *Bioresour. Technol.*, 2017, **244**, 569–574.
19. A. P. Koskina, Yu. V. Laricheva, A. I. Lysikova, O. N. Primachenko and S. S. Ivancheva, *Kinet. Catal.*, 2017, **58**, 655–662.
20. R. Liu, J. Chen, X. Huang, L. Chen, L. Ma and X. Li, *Green Chem.*, 2013, **15**, 2895–2903.
21. M. Zhang, C. Li, W. Hua, Y. Yue and Z. Gao, *Chin. J. Catal.*, 2014, **35**, 1874–1882.
22. E. Castillejos, A. M. García-Minguillán, B. Bachiller-Baeza and A. Guerrero-Ruiz, *Catal. Today*, 2018, **301**, 248–257.
23. F. Li, G. Zhang and Y. Song, *Energy Technol.*, 2018, **6**, 1633–1641.
24. G. Pagona, N. Tagmatarchis, J. Fan, M. Yudasaka and S. Iijima, *Chem. Mater.*, 2006, **18**, 3918–3920.
25. N. Tagmatarchis, A. Maigne, M. Yudasaka and S. Iijima, *Small*, 2006, **2**, 490–494.
26. C. Poonjarernsilp, N. Sano, N. Sawangpanich, T. Charinpanitkul and H. Tamon, *Green Chem.*, 2014, **16**, 4936–4943.
27. J. P. Paraknowitsch and A. Thomas, *Energy Environ. Sci.*, 2013, **6**, 2839–2855.
28. A. Sharma, K. Dasgupta, A. Patwardhan and J. Joshi, *Chem. Eng. Sci.*, 2017, **170**, 756–766.
29. X. Chen, W.-D. Oh, Z.-T. Hu, Y.-M. Sun, R. D. Webster, S.-Z. Li and T.-T. Lim, *Appl. Catal., B*, 2018, **225**, 243–257.
30. J. Li, S. Wang, H.-Y. Liu, H.-J. Zhou and Y. Fu, *ChemistrySelect*, 2017, **2**, 33–41.
31. O. Guellati, F. Antoni, M. Guerioune and D. Bégin, *Catal. Today*, 2018, **301**, 164–171.
32. S. Liu, G. Li, Y. Gao, Z. Xiao, J. Zhang, Q. Wang, X. Zhang and L. Wang, *Catal. Sci. Technol.*, 2017, 7, 4007.
33. R. Tenne, *Adv. Mater.*, 1995, 7, 965–995.
34. Y. Marco, L. Roldán, E. Muñoz and E. García-Bordejé, *ChemSusChem*, 2014, 7, 2496–2504.

35. S. Louisia, R. C. Contreras, M. Heitzmann, M. Rosa Axet, P.-A. Jacques and P. Serp, *Catal. Commun.*, 2018, **109**, 65–70.
36. M. Breysse, E. Furimsky, S. Kasztelan, M. Lacroix and G. Perot, *Catal. Rev.: Sci. Eng.*, 2002, **44**, 651–735.
37. I. Bychko, A. Abakumov, N. Lemesh and P. Strizhak, *ChemCatChem*, 2017, **9**, 4470–4474.
38. F. D. Pileidis and M.-M. Titirici, *ChemSusChem*, 2016, **9**, 562–582.
39. M. Nath, B. C. Satishkumar, A. Govindaraj, C. P. Vinod and C. N. R. Rao, *Chem. Phys. Lett.*, 2000, **322**, 333–340.
40. R. Liu, F. Li, C. Chen, Q. Song, N. Zhao and F. Xiao, *Catal. Sci. Technol.*, 2017, 7, 1217–1226.
41. I. Shimoyama and Y. Baba, *Carbon*, 2016, **98**, 115–125.

CHAPTER 4

Stability of Carbon Nanosupports

The carbon nanosupports used for the preparation of catalysts for HYD and HPR applications must exhibit sufficient stability and/or low reactivity under various pressures of H_2 and temperatures. In addition, water can be present in reaction streams either as part of the feed or as the product of HDO reactions. The CO_2 formed during the decarboxylation of some biomass-derived reactants may also react with carbon. As some steps during catalyst preparation have been conducted in air, the resistance of carbon nanomaterials to oxidation also deserves attention.

The reactivity of the surface of carbon nanomaterials alone may be different from that of the bare surface of the corresponding catalysts. Obviously, the reactivity of latter increases after anchoring of active metals because of their catalytic effects on the reactions of carbon with H_2, H_2O, CO_2 and air. Hence the stability issues regarding the reactivity of carbon nanomaterials alone and particularly in the presence of active metals may require further investigation.

4.1 Stability in Hydrogen

The long-term performance of carbon-supported catalysts under HYD conditions may be related to their resistance to any structural changes caused by H_2. Of particular importance is the resistance to the reaction known as methanation, in which the structure of carbon nanosupports is disrupted by the conversion of carbon to methane:

$$C + 2H_2 \rightarrow CH_4$$

For example, for AC and carbon nanofilaments, it was found that the methanation reaction became evident at temperatures above 450 and 550 °C,

Catalysis Series No. 37
Carbon Nanomaterials in Hydrogenation Catalysis
By Edward Furimsky

Published by the Royal Society of Chemistry, www.rsc.org

respectively.[1] This confirmed that highly crystalline carbon nanosupports are much more resistant to methanation than relatively amorphous AC. Apparently, the remarkable activity of catalysts supported on either CNTs or graphene-derived supports allows both HYD and HPR operations to be conducted under milder conditions, far remote from the methanation range. Nevertheless, even for extra-heavy feeds, the HPR temperature rarely approached the critical temperature of 450 °C, indicating the high resistance of carbon nanosupports to methanation under conditions generally employed during HYD and HPR operations. However, this situation may change in the presence of active metals, which have been observed to be catalytically active during methanation.[2,3] In that case, the long-term stability of carbon supports in H_2, particularly when they are part of catalysts, should be evaluated in more detail.

Under a certain H_2 pressure, graphene can be converted to graphane, a form of hydrogenated graphene. In this process, the sp^2 configuration of carbon bonds in graphene is changed to sp^3 in graphane.[4] However, at the temperatures and H_2 pressures employed in HYD, graphene is resistant to methanation. Therefore, for graphene, this reaction requires a temperature of at least 550 °C to proceed.[4] It appears that fullerenes are less resistant than graphene to methanation. During HYD in the presence of noble metals and some transition metals, C_{60} could be converted to $C_{60}H_{36}$. Without active metals, the methanation of C_{60} and C_{70} began at 420 °C. This suggests that the choice of fullerenes as catalyst supports for HYD and HPR applications should be carefully evaluated.

4.2 Stability in H_2O and CO_2

Their hydrophobicity makes carbon nanomaterials suitable supports for catalysts used in an aqueous environment. In this case, water is either the reaction medium or is produced during HDO reactions. The loss of carbon due to the steam gasification reaction, *i.e.*

$$C + H_2O \rightarrow CO + H_2$$

under HPR conditions was not observed even for amorphous carbons such as AC and CB.[5] Therefore, even a higher stability of nanocarbons under aqueous conditions may be expected because of their high crystallinity. Therefore, in order to proceed, the steam gasification of carbon requires temperatures higher than 700 °C. Similarly, the Boudart reaction:

$$C + CO_2 \rightarrow 2CO$$

requires temperatures higher than 700 °C to proceed.[6] Rodriguez *et al.*[7] reported that highly crystalline CNFs began to gasify in CO_2 at 850 °C compared with about 600 °C for amorphous carbon solids. Even in air, a temperature of 640 °C was necessary to observe the decomposition of CNTs.[8]

These temperatures are remote from the temperature ranges encountered in HYD and HPR reactions. This situation may change if active metals are anchored on the carbon supports. For example, some transition metals are well-known catalysts of the steam and CO_2 gasification reactions mentioned above.[9] Another issue that deserves attention is the loss of metals due to leachability.[10,11]

4.3 Stability in Air

During the preparation of catalysts supported on carbon nanomaterials, various steps (*e.g.* drying, tempering) may be conducted in the presence of air. Therefore, the stability of carbon nanosupports in air is of interest. In this regard, Guo *et al.*[12] compared two types of CNTs, graphene nanoplatelets (GNPs) and CB, during temperature-programmed oxidation (TPO). Some properties of these materials are given in Table 4.1. During TPO, the temperature was increased to 950 °C at 10 °C min^{-1} in a flow of air. Figure 4.1[12] shows that for CNTs and CB, a temperature of 500 °C was required to see any evidence of oxidation. For GNPs, a minor weight loss between 300 and 400 °C is consistent with the burning of some impurities rather than the graphene material. In the major region of combustion (above 600 °C), the burn-off rate of graphene, indicated by the slope of the curve, was the lowest. At the same time, rapid weight loss after ignition indicates the amorphous nature of CB. Both CNTs exhibited almost identical resistance to oxidation by air. Interestingly, little oxygen chemisorption was observed below 500 °C. This contrasts with the weight increase due to oxygen chemisorption during similar TPO reactions using various carbonaceous solids.[13] For these solids, the amount of chemisorbed oxygen increased with increase in the H : C ratio of the solids, suggesting that the C–H entities were involved by facilitating the formation of hydrogen peroxy radicals. However, because of the absence of hydrogen in the carbon nanomaterials, a similar weight increase could not be observed.

Some metals that are part of HYD and HPR catalysts supported on carbon nanomaterials are catalytically active during carbon oxidation reactions.[3] Therefore, the stability of these catalysts in air is of concern. In fact, from the health and safety point of view, these catalysts should be classified as pyrophoric solids, requiring special handling procedures designed for such solids.[10,11]

Table 4.1 Properties of carbon nanomaterials.[12]

Carbon nanomaterial	Surface area per $m^2 g^{-1}$	Pore volume per $cm^3 g^{-1}$	Raman I_D/I_G
CB	1339	1.71	1.87
GNPs	63	0.14	0.28
CNT1	114	0.21	0.84
CNT2	58	0.09	0.44

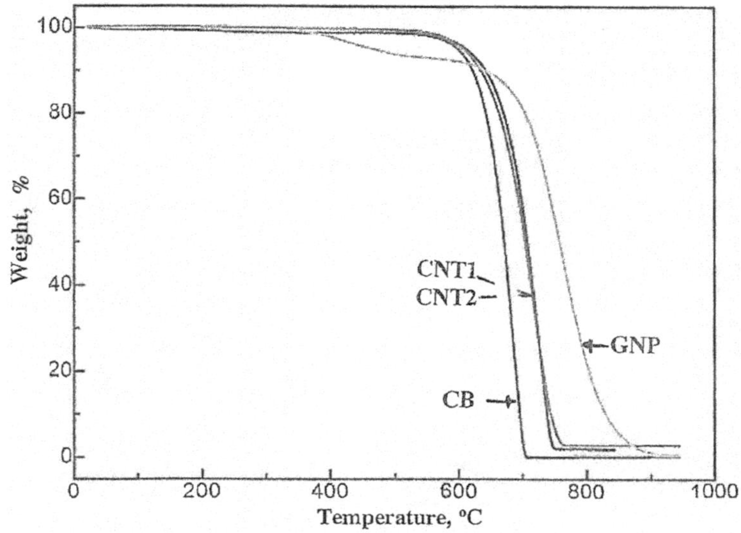

Figure 4.1 TPO profiles of the solids in Table 4.1.
Reproduced from ref. 12 with permission from John Wiley and Sons, © 2017 Wiley-VCH Verlag GmbH & Co. KGaA, Weinheim.

4.4 Stability in Liquid Media

There should be little concern about the stability of carbon nanomaterials in polar and nonpolar liquids and also the solvents used during the evaluation of various reactants under HYD conditions.[3] Also, the highly crystalline carbons are known for high stability in reaction streams used in HPR.[2] Modification of the structure of carbon nanomaterials by strong acids such as those used for functionalization was discussed earlier. Therefore, it is evident that, in comparison with the most frequently used supports of HPR catalysts, such as γ-Al_2O_3, carbon nanomaterials are much more stable. As part of catalysts, the potential leaching of active metals deposited on carbon supports may require attention.

References

1. L. Li, C. Liang, Z. Feng, P. Ying, D. Wang and C. Li, *J. Mol. Catal. A: Chem.*, 2004, **211**, 103–109.
2. E. Furimsky, *Carbons and Carbon Supported Catalysts in Hydroprocessing*, RSC Publishing, Cambridge, UK, 2008.
3. P. Serp, M. Corrias and P. Kalck, *Appl. Catal.*, 2003, **253**, 337–358.
4. D. C. Elias, R. R. Nair, T. M. G. Mohiuddin, S. V. Morozov, P. Blake, M. P. Halsall, A. C. Ferrari, D. W. Boukhvalov, M. I. Katsnelson, A. K. Geim and K. S. Novoselov, *Science*, 2009, **323**, 610–613.
5. J. Albero and H. Garcia, *J. Mol. Catal. A: Chem.*, 2015, **408**, 296–309.

6. E. Furimsky, *Appl. Catal.*, 1987, **33**, 97–106.
7. N. M. Rodriguez, A. Chambers and R. T. K. Baker, *Langmuir*, 1995, **11**, 3862–3866.
8. T. T. Nguyen and P. Serp, *ChemCatChem*, 2013, **5**, 3595–3603.
9. E. Furimsky, *Energy Fuels*, 2008, **22**, 237–242.
10. E. Furimsky, *Ind. Eng. Chem. Res.*, 2013, **52**, 17695–17713.
11. M. Marafi and E. Furimsky, *Energy Fuels*, 2017, **31**, 5711–5750.
12. K. Guo, M. Gu and Z. Yu, *Energy Technol.*, 2017, **5**, 1228–1234.
13. E. Furimsky, *Fuel Process. Technol.*, 1988, **19**, 203–210.

CHAPTER 5

Reactants and Feeds

The investigation of reactants and feeds with widely varying compositions and origins was prompted by the requirements for high activity and unique selectivity that can be achieved over catalysts supported on carbon nanomaterials.[1] This included the most complex feeds derived from biomass. The reactants typical of those present in conventional petroleum, coal-derived liquids and liquids obtained from oil shale have also been included, although to a lesser extent. Because of environmental and health concerns, halogen-containing reactants have been studied.

5.1 Biomass-derived Reactants

The reactants derived from biomass have been identified as a potential source of fine chemicals and also fuels and/or blending components with fuels. Much attention has been devoted to α,β-unsaturated C=C and C=O conjugated double-bonded aldehydes. Similarly, the components of lignocellulosic biomass such as cellulose, hemicellulose and lignin have been evaluated as a renewable source of value-added products. For example, several useful products and/or intermediates can be identified in the mechanism of the depolymerization of cellulose and hemicellulose (Figure 5.1) developed by Pileidis and Titirici.[2] However, in the presence of a catalyst, cellulose can be converted to different products. For example, at 513 K and 4 MPa of H_2 in a batch reactor over monometallic Pd and bimetallic Pd–Fe catalysts supported on CNTs, cellulose was converted to polyols such as mannitol, ethylene glycol, 1,2-propanediol and glycerol.[3] For the bimetallic catalyst, the addition of Fe resulted in an enhancement of selectivity to mannitol, but losses of Fe due to leaching affected the selectivity. Important chemicals and a high-quality biodiesel fuel can be prepared from some components present in vegetable oil and algae biomass.

Catalysis Series No. 37
Carbon Nanomaterials in Hydrogenation Catalysis
By Edward Furimsky

Published by the Royal Society of Chemistry, www.rsc.org

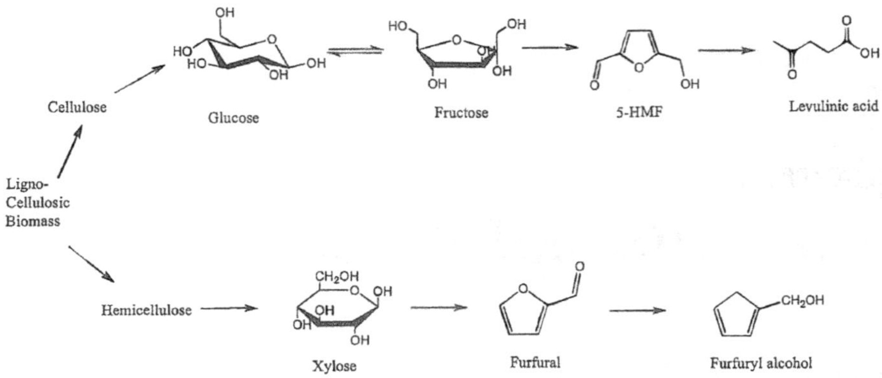

Figure 5.1 Mechanisms of depolymerization of cellulose and hemicellulose.

5.1.1 α,β-Unsaturated Reactants

α,β-Unsaturated aldehydes are suitable reactants for studying the structure-sensitive HYD of conjugated C=C and C=O double bonds. Among such reactants, cinnamaldehyde (CALD) and citral have attracted attention because of their industrial importance. For these reactants, the preferential HYD of the C=O bond, which is part of conjugated C=O and C=C structures, is desirable to obtain products and/or intermediates of interest.[4] Thus, thermodynamically, there is a greater driving force for the HYD of C=C bonds than for the HYD of C=O bonds.[5] The preferential HYD of C=O bonds can be achieved experimentally provided that a special geometry of the transition state, involving active sites and adsorbed reactant on the surface of the catalyst, can be established. Then, fairly mild conditions are sufficient to obtain high yields of the targeted products. Otherwise, complete HDO, followed by HYD of the α,β-aldehydes to hydrocarbons, is achieved when the experimental conditions approach those employed during HPR.

5.1.1.1 Cinnamaldehyde

CALD occurs naturally as *trans*-CALD and can be isolated from cinnamon essential oil. It can also be synthesized in the laboratory. Structurally, it consists of a benzene ring attached to an unsaturated aldehyde. In the HYD of CALD (Figure 5.2), either cinnamyl alcohol (CALC) or hydrocinnamaldehyde (HCALD) can be produced. Both are important intermediates for the industrial production of fine chemicals such as perfumes and pharmaceuticals. The HYD of both C=C and C=O double bonds leads to the formation of hydrocinnamyl alcohol (HCALC). Obviously, if more severe conditions are employed, complete HDO leading to hydrocarbon formation is observed.

Because of the three double bonds in unsaturated aldehydes, *i.e.* isolated benzene ring C=C and conjugated C=O and C=C double bonds, the selection of the catalysts and experimental conditions requires special

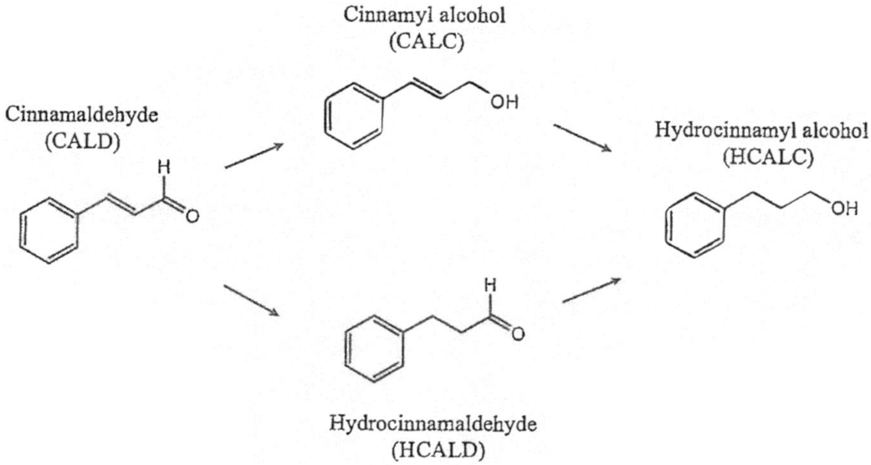

Figure 5.2 Tentative mechanism of conversion of cinnamaldehyde.

attention. Thus, if CALC is the product of interest, the more reactive C=C bond must be protected in favor of the selective HYD of the C=O bond. Otherwise, undesirable products such as saturated alcohols and aldehydes, *i.e.* HCALC and HCALD, respectively, would be obtained. Therefore, catalysts with high selectivity with respect to the HYD of the C=O bond to a C–OH entity, while leaving the C=C bond intact, are required. Catalysts supported on carbon nanomaterials have been investigated for these applications. In this regard, the electronic effects of nanocarbon supports, combined with active metals of an appropriate type, and also the size and morphology of the active metal particles deposited on the carbon supports play an important role in selective HYD catalysis involving α,β-unsaturated aldehydes as reactants.[6]

5.1.1.2 *Citral*

In Nature, citral occurs in lemon plants. Citral is also an α,β-conjugated unsaturated aldehyde, and occurs in the form of double-bond isomers, *i.e.* the *E*- and *Z*-isomers known as geranial and neral, respectively, and also as citral A and citral B, respectively. The HYD products are referred to as geraniol and nerol. Generally, the ratio of geraniol to nerol in HYD products corresponds to the ratio of geranial to neral in the reactant mixture. Because of its lemon or citrus odor, citral is an aroma compound used in perfumery and as a flavor, and also for fortifying lemon oil.[7] Citral is also used in the synthesis of vitamin A, ionone, methylionone and other products.

Figure 5.3 shows a tentative mechanism of the HYD of citral.[8] Similarly to the HYD of CALD, the objective is to hydrogenate the C=O bond while protecting the C=C bond. Thus, geraniol and nerol are frequently the targeted products. To a lesser extent, citronellal and citronellol are also sought.

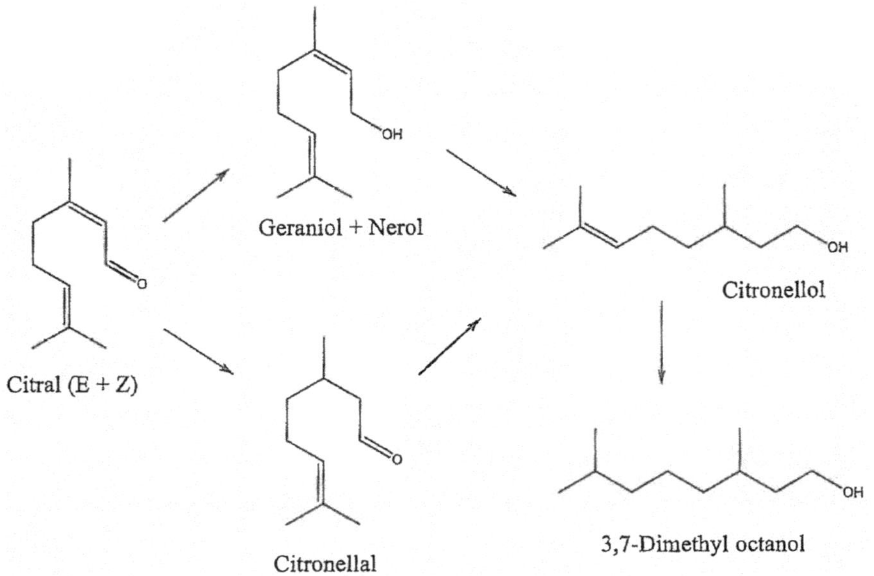

Figure 5.3 Tentative mechanism of conversion of citral.

Similarly to the HYD of CALD, catalysts exhibiting high selectivity for the HYD of the C=O bond in α,β-unsaturated aldehydes are required for these applications.[9] Otherwise, under less controlled conditions, HYD may continue to reach the other products shown in Figure 5.3. Eventually, conditions for complete HDO can be reached to produce a high-octane C_{10} isoalkane, a suitable blending component in gasoline. The separation of isomers (*e.g.* geraniol from nerol) during the HYD of citral is also of interest.

5.1.1.3 *Crotonaldehyde*

Crotonaldehyde occurs in Nature and can also be synthesized from acetaldehyde by aldol condensation. Its main use is as a reactant for the preparation of sorbic acid, which is a food preservative, crotonic acid, which is a precursor of vitamin E, 3-methoxybutanol, *etc.*[10] Of particular importance is 3-methylcrotonaldehyde (3-methyl-2-butenal; prenal). The selective HYD of prenal leads to the formation of the unsaturated alcohol prenol, which is a starting reactant in chemical industry for the preparation of various chemicals such as fragrances, flavorings and pharmaceuticals. In this regard, high activity and selectivity of catalysts supported on carbon nanomaterials are required.

5.1.2 Cellulose- and Hemicellulose-derived Reactants

As the most abundant renewable source, lignocellulosic biomass can be depolymerized to cellulose and hemicellulose, which, in the next stage, can be converted to glucose and fructose and to xylose, respectively (Figure 5.1).[11] In this case, either hydrolysis or selective catalytic HYD can be applied.[12] The primary products from depolymerization can be further converted to a wide range of useful products, *e.g.* sorbitol, mannitol, furfural and levulinic acid. During final upgrading, the yield of the products of interest is influenced by the activity and selectivity of the catalyst employed.

5.1.2.1 Conversion of Sorbitol

Based on the above information, sorbitol is part of the overall catalytic transformation of cellulose to hydrocarbons, *i.e.* cellulose → glucose → sorbitol → hydrocarbons. Otherwise, high yields of sorbitol can be obtained by the HYD of glucose under optimized conditions.[12] In this case, selective reduction of the aldehyde group of D-glucose to a hydroxyl group is involved. Complete HDO of sorbitol results in the formation of C_5 and C_6 hydrocarbons suitable for blending with gasoline. Interest in this route has resulted from lignocellulosic biomass being one of the most abundant types of biomass.

5.1.2.2 Conversion of Furans and Furfural

2-Methylfuran (2MF) and 2,5-dimethylfuran (DMF) are biomass-derived compounds used for the production of perfume intermediates, pesticides, pharmaceutical intermediates, *etc.* Value-added products can also be obtained *via* HYD and/or hydrogenolysis of the furan ring, yielding commercial products such as THFs and alcohols.

Furfural can be obtained by the gradual transformation of a lignocellulosic component such as hemicellulose *via* xylose, as shown in Figure 5.1. The selective HDO of the carbonyl group of furfural to obtain a methyl group gives methylfuran, a fuel blending component with high octane number (*e.g.* RON ≈ 131), high energy density and low water solubility. Similarly, 5-hydroxymethylfurfural (5-HMF) can be prepared by the hydrolysis of glucose.[13] A good catalyst for these applications must selectively hydrogenate the carbonyl C=O group while leaving C–O, C–C and C=C entities of the furan ring intact.

The selective HYD of furfural to furfuryl alcohol (FFA) is of industrial importance. FFA is a feed for the preparation of green solvents such as tetrahydrofurfuryl alcohol (THFFA) and also resins, synthetic fibers, perfumes and vitamins. Also, THFFA is a reactant for the preparation of 1,5-pentanediols, printing inks, agricultural products and electronics cleaners.

5.1.2.3 Levulinic Acid

Levulinic acid (LA) can be readily prepared by the acid hydrolysis (*e.g.* with HCl or H_2SO_4) of lignocellulosic components such as glucose, fructose and starch.[14] In this reaction, 5-HMF is the intermediate before LA is obtained, as indicated in Figure 5.1. It is the feed for the preparation of γ-valerolactone (GVL) *via* selective HYD. In this case, the C=O group is selectively hydrogenated to an OH group to give 4-hydroxyvaleric acid (4-hydroxypentanoic acid), which is subsequently dehydrated to obtain GVL, as shown in Figure 5.4. GVL can be used in several industrial applications,[15,16] for example, as a solvent and blending component of gasoline. Also, after the HYD of GVL, alkanes and alkenes suitable for blending with gasoline fuel can be obtained.[17] In addition, more than a dozen products arising from LA were described by Pileidis and Titirici.[2]

5.1.3 Lignin and Monomeric Phenols

In addition to cellulose and hemicellulose, lignin (Figure 5.5) is another component obtained by the depolymerization of lignocellulosic biomass. Together with hemicellulose, lignin is separated from cellulose during paper manufacture and is therefore available in large quantities as a low-cost material. Lignin is a three-dimensional amorphous polymer consisting of repeatedly, but randomly, cross-linked phenol units. Compared with cellulose and hemicellulose, lignin is much more resistant to depolymerization to give monomeric aromatics and phenols. Among potential methods, the catalytic hydrogenolysis of lignin has attracted attention.[18,19] In this regard, the breakdown of the abundant β-O-4 linkages (Figure 5.5) is of primary importance before aromatic and phenolic monomeric products can be

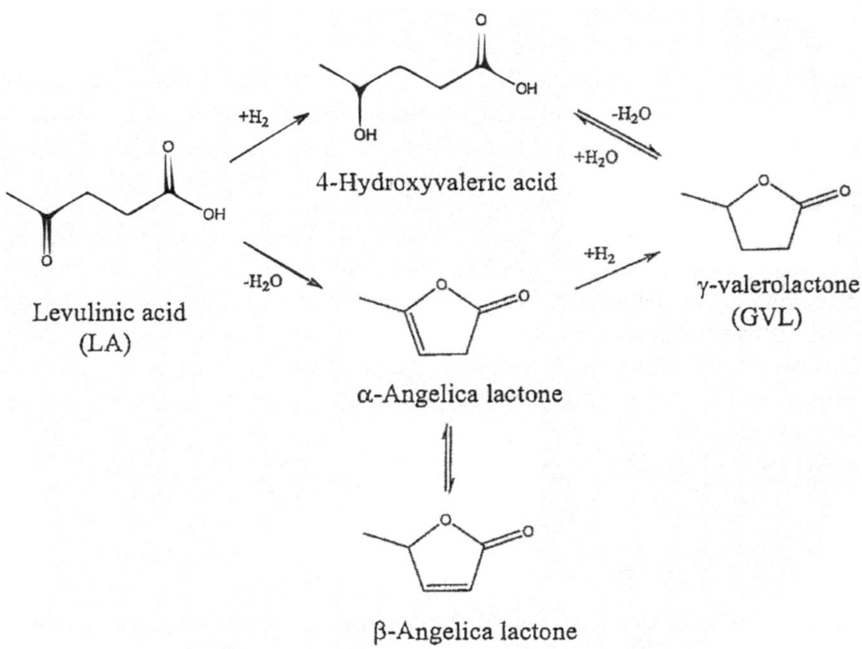

Figure 5.4 Tentative mechanism of conversion of levulinic acid.

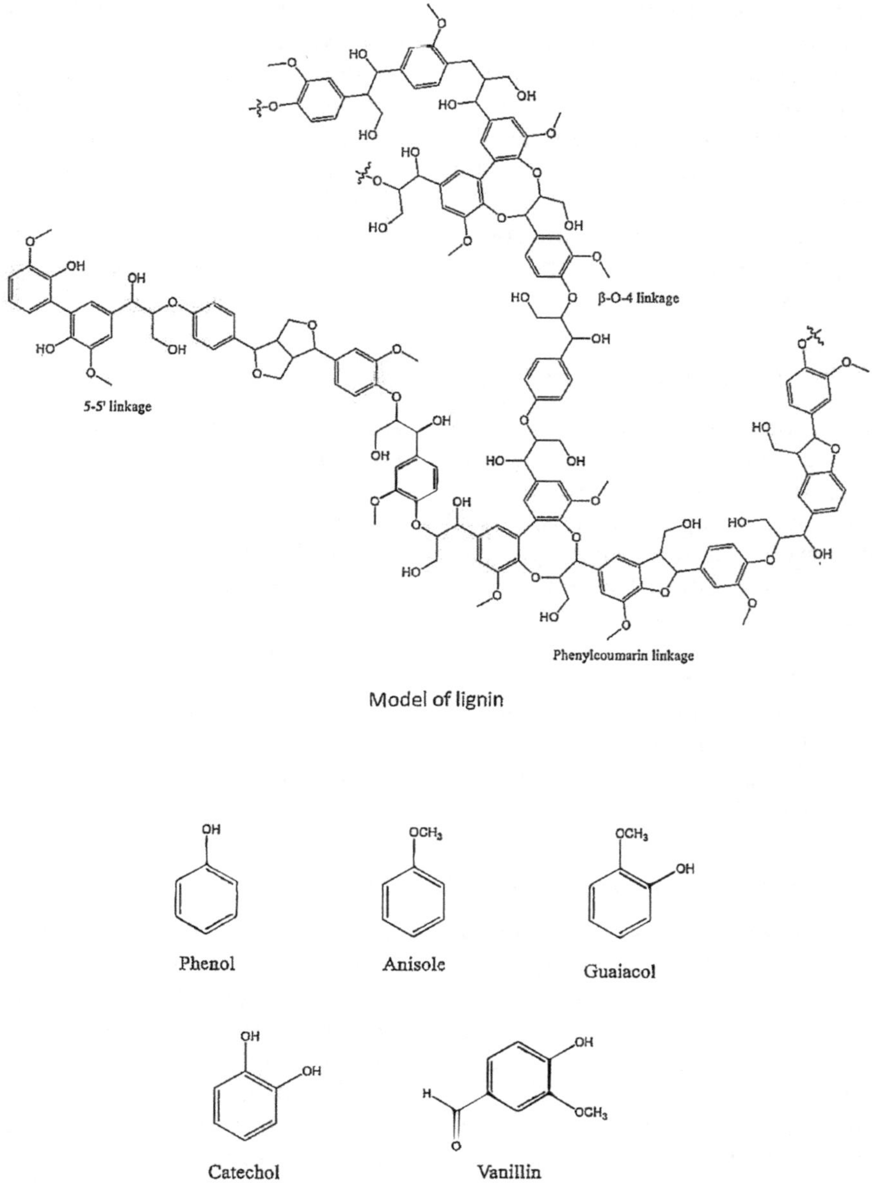

Figure 5.5 Model of lignin and basic monomers derived from lignin.

released.[20] Other linkages that can be found in the structure of lignin were discussed in detail by Chakar and Ragauskas.[21]

Monomeric phenols are major products of lignin depolymerization. For example, in the lignin depolymerization study conducted by Patil *et al.*,[22] the liquid product consisted mainly of substituted phenols (*e.g.* guaiacols, syringols, eugenols) as major components. The structures of some monomeric phenols are shown in Figure 5.5. Under HPR conditions, cleavage of the C–O bond results in the formation of OH groups, which are eliminated *via* dehydration to form C=C bonds before the final HYD step leading to the formation of hydrocarbons. Under milder conditions such as generally employed in HYD, monomeric phenolic structures are the predominant species in the products. These products can be converted to hydrocarbons under HPR conditions in the next stage.[11] Such hydrocarbons are suitable as blending components of transportation fuels. Monoaromatics and phenols have also been used as starting materials for the production of chemicals, polymers, *etc.*

5.1.4 Triglycerides and Fatty Acids

Triglycerides are the predominant components of vegetable oil biomass, but various concentrations of free fatty acids, monoglycerides, diglycerides and some lipids may also be present. Both edible and inedible vegetable oils and fats can be found. The former include vegetable oils from soy, canola, sunflower, safflower, peanut, cottonseed, *etc.*, while linseed oil, tung oil and castor oil are typical examples of inedible oils.[23] Methods such as chemical extraction, mechanical extraction and expeller pressing extraction have been used for the recovery of vegetable oils from plant seeds and other parts of plants.[24,25] As a biofeed, vegetable oils can be converted to primary products using either hydrothermal liquefaction or pyrolysis. The products obtained need further upgrading *via* HPR to obtain green diesel fuel.[26] A similar methodology can be employed to upgrade algae biomass containing large amounts of lipids.

5.2 Petroleum-derived Reactants

Decades of commercial utilization of conventional petroleum suggest that the structures of the components have been thoroughly elucidated and are well documented in the literature, including novel feeds such as those derived from light tight oils and oil shale. The focus has been on S- and N-containing compounds and also on aromatics, in response to environmental regulations. The conversion of *n*-alkanes to isoalkanes was required to attain the performance parameters of fuels and lubricants. Rather than repeating this material here, reference is made to authoritative reviews that can be readily accessed.[27–29] Catalysts supported on carbon nanosupports have also been evaluated for the HPR of conventional feeds. As indicated earlier, the test conditions differed from those used in HYD studies. The selective HYD of styrene to ethylbenzene is necessary to achieve stability of pyrolysis gasoline. Gaseous hydrocarbons produced by pyrolysis contain acetylene, which can be converted to ethylene, an important monomer in the plastics industry. The interest in catalysts supported on carbon nanomaterials, exhibiting high selectivity for these reactions, has been noted.[30,31]

5.3 Other Reactants

The performance of carbon nanomaterial-supported catalysts, with unique selectivity during reactions of interest in particular, suggests that these catalysts may find applications in various environmentally related methods. So far, only AC-supported catalysts have been used.[32] The advantage of AC-supported catalysts over those supported on oxidic supports was clearly evident.[33,34] The available information indicates that some carbon nanosupports have much greater beneficial effects on catalyst performance than AC supports.[35–39]

Catalytic treatment of hazardous waste appears to be cost competitive with currently used incineration methods.[40] For example, the hazardous nature of chlorine-containing organic solid and/or liquid wastes can be minimized by catalytic treatment, and hydrodechlorination (HDCl) reactions over noble metal-containing catalysts have attracted attention.[41,42] Typical feeds used for HDCl differ markedly in their composition from conventional and well-established nonconventional feeds. They comprise polychlorinated biphenyls (PCBs), waste chlorinated polymers, *e.g.* poly(vinyl chloride) (PVC), some water streams contaminated by Cl-containing species and waste chemicals. It was shown that PCBs can be converted under fairly mild conditions to benign products over catalysts supported on rGO.[36] Other materials of environmental concern are chlorophenols, which are widely used in several industrial applications, *e.g.* as herbicides, insecticides, wood preservatives and disinfectants.[43] Therefore, catalytic HDCl may play a significant role in various environmental applications.

The incineration of waste plastics containing nitrogen may lead to the formation of very toxic species such as HCN.[44] This can be avoided in the presence of a catalyst. In fact, under certain conditions, nitrogen in the waste can be converted to N_2.[45] The potential of carbon nanosupports in such applications has not yet been addressed.

References

1. J. Albero and H. Garcia, *J. Mol. Catal. A: Chem.*, 2015, **408**, 296–309.
2. F. D. Pileidis and M.-M. Titirici, *ChemSusChem*, 2016, **9**, 562–582.
3. M. Nath, B. C. Satishkumar, A. Govindaraj, C. P. Vinod and C. N. R. Rao, *Chem. Phys. Lett.*, 2000, **322**, 333–340.
4. F. Delbecq and P. Sautet, *J. Catal.*, 1995, **152**, 217–236.
5. P. Gallezot and D. Richard, *Catal. Rev.: Sci. Eng.*, 1998, **40**, 81–126.
6. K. Inuzuka, *Bull. Chem. Soc. Jpn.*, 1961, **34**, 1557–1560.
7. *citral* in *The Merck Index*, 12th edn.
8. E. Bailón-García, F. J. Maldonado-Hódar, A. F. Pérez-Cadenas and F. Carrasco-Marín, *Catalysts*, 2013, **3**, 853–877.
9. A. Stolle, T. Gallert, C. Schmoger and B. Ondruschka, *RSC Adv.*, 2013, **3**, 2112–2153.

10. R. P. Schulz, J. Blumenstein and C. Kohlpaintner, *Crotonaldehyde and Crotonic Acid, Ullmann's Encyclopedia of Chemical Technology*, Wiley-VCH, Weinheim, 2005.
11. R. Rinaldi and F. Schueth, *ChemSusChem*, 2009, **2**, 1096–1107.
12. D. Klemm, B. Heublein, H.-P. Fink and A. Bohn, *Angew. Chem., Int. Ed.*, 2005, **44**, 3358–3393.
13. L. Liu, H.-m. Chang, H. Jameel, J.-Y. Park and S. Park, *Ind. Eng. Chem. Res.*, 2017, **56**, 14447–14453.
14. I. T. Horváth, H. Mehdi, V. Fábos, L. Boda and L. T. Mika, *Green Chem.*, 2008, **10**, 238–242.
15. J. Q. Bond, D. M. Alonso, D. Wang, R. M. West and J. A. Dumesic, *Science*, 2010, **327**(596), 1110–1114.
16. L. Yan, Q. Yao and Y. Fu, *Green Chem.*, 2017, **19**, 5527–5547.
17. X. Tang, X. Zeng, Z. Li, L. Hu, Y. Sun, S. Liu, T. Lei and L. Lin, *Renewable Sustainable Energy Rev.*, 2014, **40**, 608–620.
18. T. Phongpreecha, N. C. Hool, R. J. Stoklosa, A. S. Klett, C. E. Foster, A. Bhalla, D. Holmes, M. C. Thies and D. B. Hodge, *Green Chem.*, 2017, **19**, 5131–5143.
19. B. N. Kuznetsov, I. G. Sudakova, N. V. Garyntseva, L. Djakovitch and C. Pinel, *React. Kin. & Catal. Lett.*, 2017, **120**, 81–94.
20. L. Li, C. Liang, Z. Feng, P. Ying, D. Wang and C. Li, *J. Mol. Catal. A: Chem.*, 2004, **211**, 103–109.
21. F. S. Chakar and A. J. Ragauskas, *Ind. Crops Prod.*, 2004, **20**, 131–141.
22. P. T. Patil, U. Armbruster, M. Richter and A. Martin, *Energy Fuels*, 2011, **25**, 4713–4722.
23. A. Thomas, *Fats and Fatty Oils. Ullmann's Encyclopedia of Industrial Chemistry*. Wiley-VCH, Weinheim, 2002.
24. R. A. Moreau, M. I. Powell and K. B. Hicks, *J. Agric. Food Chem.*, 1996, **44**, 2149–2154.
25. M. Eisenmenger, N. T. Dunford, F. Eller, S. Taylor and J. Martinez, *J. Am. Oil Chem. Soc.*, 2006, **83**, 863–868.
26. L. Li, E. Coppola, J. Rine, J. L. Miller and D. Walker, *Energy Fuels*, 2010, **24**, 1305–1315.
27. H. Topsoe, B. S. Clausen and F. E. Massoth, Hydrotreating Catalysis, in *Catalysis-Science and Technology*, ed. J. Anderson and M. Boudart, 1996, vol. 11, pp. 1–310.
28. F. E. Massoth, *Adv. Catal.*, 1978, **27**, 265–290.
29. T. Ho, *Catal. Rev.:Sci. Eng.*, 1988, **30**, 117–154.
30. Z. Liu, Z. Li, F. Wang, J. Liu, J. Ji, K. C. Park and M. Endo, *Mater. Res. Bull.*, 2012, **47**, 338–343.
31. V. V. Chesnokov, O. Y. Podyacheva and R. M. Richards, *Mater. Res. Bull.*, 2017, **88**, 78–84.
32. M. Marafi and E. Furimsky, *Energy Fuels*, 2017, **31**, 5711–5750.
33. E. Furimsky, *Ind. Eng. Chem. Res.*, 2013, **52**, 17695–17713.
34. E. Furimsky, *Catal. Today*, 2013, **217**, 13–56.

35. H. Li, A. Riisager, S. Saravanamurugan, A. Pandey, R. S. Sangwan, S. Yang and R. Luque, *ACS Catal.*, 2018, **8**, 148–187.
36. X. Guo, C. Yu, Z. Yin, S. Sun and C. T. Seto, *ChemSusChem*, 2018, **11**, 1617–1620.
37. L. Rodríguez-Perez, M. Ángeles Herranza and N. Martín, *Chem. Commun.*, 2013, **49**, 3721–3735.
38. N. G. Sahoo, Y. Z. Pan, L. Li and S. H. Chan, *Adv. Mater.*, 2012, **24**, 4203–4210.
39. E. Furimsky, *Ind. Eng. Chem. Res.*, 2017, **56**, 11359–11371.
40. M. Marafi, A. Stanislaus and E. Furimsky, *Handbook of Spent Hydroprocessing Catalysts*, Elsevier, Amsterdam, 2nd edn, 2016.
41. M. Martin-Martinez, A. Álvarez-Montero, L. M. Gómez-Sainero, R. T. Baker, J. Palomar, S. Omar, S. Eser and J. J. Rodriguez, *Appl. Catal., B*, 2015, **162**, 532–543.
42. M. Martin-Martinez, M. Gómez-Sainero, J. Palomar, S. Omar and J. J. Rodriguez, *Catal. Lett.*, 2016, **146**, 2614–2621.
43. W. Chang, H. Kim, G. Y. Lee and B. J. Ahn, *Res. Chem. Intermed.*, 2016, **42**, 71–82.
44. M. Nielsen, P. Jurasek, J. Hayashi and E. Furimsky, *J. Anal. Appl. Pyrolysis*, 1995, **35**, 43–51.
45. E. Furimsky, M. Nielsen and P. Jurasek, *Energy Fuels*, 1995, **9**, 439–447.

CHAPTER 6

Development of Catalysts Supported on Carbon Nanosupports

Attempts have been made to develop catalysts with desirable activity and selectivity for the products of interest in various HYD and HPR applications. Carbon nanomaterials derived from graphene (*e.g.* GO and rGO), carbon nanotubes (CNTs), carbon nanofibers (CNFs), carbon nanohorns (CNHs) and fullerenes have been evaluated as potential supports. In several studies, carbon blacks (CBs) and activated carbons (ACs) together with conventional HPR catalysts have been used for comparison. Noble metals (*e.g.* Pt, Pd, Rh, Ir and Ru) and transition metals (*e.g.* Mo, W, Co, Ni, Fe and Cu) and also rare earth metals (*e.g.* Y, La and Ce) have been evaluated as catalytically active metals. Frequently, the catalysts supported on carbon nanomaterials outperformed other catalysts.

For testing the activity and selectivity of carbon nanomaterial-supported catalysts, reactants and feeds varying widely in structure and origin have been included. Biofeeds and reactants derived from biomass have attracted considerable attention. To a lesser extent, feeds and reactants derived from conventional and nonconventional petroleum have also been used. Most of the work has been performed in batch reactors, and continuous systems have been utilized less frequently.

6.1 Activity of Carbon Nanosupports

To various extents, both supports and active metals influence the course of catalytic reactions. Decoupling of these two contributions to the overall catalysis is complex task. First, an examination of carbon nanosupports,

Catalysis Series No. 37
Carbon Nanomaterials in Hydrogenation Catalysis
By Edward Furimsky

Published by the Royal Society of Chemistry, www.rsc.org

particularly their activity in the absence of active metals, may provide valuable information. With respect to the objectives of this book, the activity of carbon nanomaterials under HYD and HPR conditions is of interest although such solids may also exhibit some catalytic activity during non-HYD reactions.[1] Information on the catalytic activity and selectivity of carbon nanosupports may be used for decoupling the fairly complex effects of the supports from those of active metals. First, in the absence of active metals, the surface of carbon supports may differ from that of carbon supports not covered with active metals, as is the case with catalysts. In the latter case, the uncovered surface of a support may serve as a reservoir of active hydrogen spilt over from the active metals on the support. The mobility of such hydrogen, after spillover from the support, has been noted.[2]

For pristine carbon nanosupports, the capability of the carbon surface to activate hydrogen is fairly low.[2] Thus, to be catalytically active in HYD reactions, carbon nanomaterials alone must be capable of activating dihydrogen, as tentatively depicted in the following scheme:

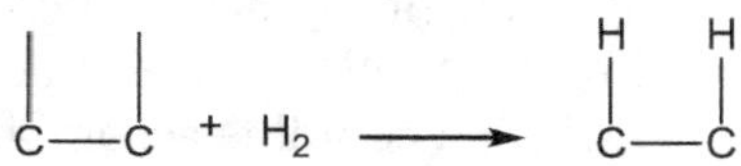

Subsequently, the active hydrogen must be readily transferred to a reactant molecule to initiate HYD reactions, as illustrated for the tentative HYD of ethene to ethane, *i.e.*

In the latter reaction, the active site for hydrogen activation is recovered for another HYD cycle. Hence these are the essential requirements for carbon nanosupports to be active in HYD reactions in the absence of catalytically active metals.[2]

Amorphous carbons such as ACs and CBs exhibited some activity for hydrogen activation because of significant irregularities in their structures. However, the highly ordered structure of pristine carbon nanomaterials is in contrast with the structure of amorphous carbon supports. Consequently, the reactivity of pristine carbon nanomaterials, particularly that of basal planes, is very low. Some reactivity is exhibited by the edge regions of CNTs, CNFs, CNHs and peripheral areas of graphene. Of course, this situation may change significantly after functionalization and/or doping of pristine carbon nanosupports.

It has been shown that the curvature of basal planes, as evident in CNTs, CNFs, CNHs and fullerenes, and the absence of such curvature in graphene may play certain role during interaction with H_2.[3,4] Such curvature may induce partial localization of π-electrons, which results in an increase in sp^3 character. This makes the interaction with hydrogen atoms stronger compared with the planar basal planes such as are present in graphene.

However, no systematic study focusing on this aspect of carbon nanosupports could be found in the literature.

6.1.1 Activity of Graphene

In the basal planes of pristine graphene, C–H entities could be created by attaching H atoms to carbons. For example, as shown by Elias *et al.*,[5] H atoms could be generated by dc plasma treatment of an Ar–H_2 mixture in the presence of graphene. Subsequently, graphene was converted to graphane. Whereas the conversion of graphene to graphane *via* the addition of atomic H can be achieved, a similar conversion using H_2 is thermodynamically unfavorable because of the high barrier for dissociative adsorption.[6] However, this barrier can be lowered by applying a perpendicular electric field.[7] Although graphane could be synthesized using molecular H_2, it required H_2 pressures in the range 2.6–5.0 GPa (2600–5000 MPa).[8] From a practical point of view, such pressures of H_2 are unrealistic. This unfavorable situation may be alleviated by introducing catalytically active metals on the surface of graphene, as demonstrated by Parambhath *et al.*,[9] according to whom the HYD of graphene deposited with Pd could be accomplished at 2 MPa of H_2. With respect to the hydrogen activation required for HYD reactions, this situation is much more realistic.

Primo *et al.*[10] reported that the activity of rGO (without active metals) during the HYD of cyclohexene and 1-octene was enhanced by applying an H_2 plasma to rGO. The activity increase was attributed to the creation of irregularities such as carbon vacancies in the rGO structure. The tests were conducted in a batch reactor at 120 °C for cyclohexene and 80 °C for 1-octene under 3 MPa of H_2 using *n*-heptane as a solvent. Under these conditions, cyclohexene was completely converted to cyclohexane whereas for 1-octene, in addition to *n*-octane, skeletal isomers were also present. This is an indication of a bifunctional character of the H_2 plasma-treated rGO. Apparently, the C–OH entities on rGO were the source of protons necessary for isomerization reactions. Little conversion of the reactants over the untreated rGO was observed.

Shaikh *et al.*[11] observed the catalytic activity of GO during the dehydration of fructose. The GO prepared from graphite oxide was purified using an HCl solution and washed with deionized water and ethanol. The GO was mixed with fructose in the absence of any solvent and sonicated at 100 °C. Under these conditions, the selective dehydration of fructose to 5-HMF was observed. This suggests that during the catalytic HYD of oxygenates, the bare surface of GO may participate during the dehydration of intermediates as part of the overall HDO. From a mechanistic point of view, such a reaction can be enhanced by protons, which can be supplied by GO.[12,13] Even rGO could catalyze such reactions in spite of the lower concentration of –OH and –COOH groups on the surface compared with GO. This was confirmed in the depolymerization of cellulose to glucose in subcritical water.[14] Apparently, H^+ ions supplied from the rGO surface by –OH and –COOH groups were involved. The reaction rate could be further increased by microwave treatment. Under the same conditions, the activity of CNTs was very low.

Wang *et al.*[15] evaluated the catalytic activity of carbon materials alone (*e.g.* ACs, CNTs, graphene, GO and rGO) in the HYD of cellobiose to sorbitol (batch reaction, 463 K, 5 MPa H_2, 3 h). Almost complete conversion of the feed was observed over every carbon solid. However, there were significant differences in the selectivity with respect to sorbitol. Thus, rGO was the most selective, giving about 25% of sorbitol. and for all carbon solids tested the following order of selectivity with respect to sorbitol was established: rGO > GO ≈ CNTs ≫ AC > G. As expected, the selectivity of pristine graphene (G) with respect to sorbitol was very low.

The activity of nitrogen-doped rGO was confirmed by Liu *et al.*[16] in the HYD of anthracene. The rGO used in this study was prepared from GO obtained from oxidized graphite.[17] For doping with nitrogen, the rGO was treated with NH_3 at 450, 650, 800, 900, 1000 and 1100 °C for 40 min. A nearly linear increase between the amount of adsorbed anthracene and graphitic N suggests that the latter facilitated the π–π interactions of anthracene with the surface. The HYD of anthracene was carried out in a fixed-bed reactor at 350 °C, 4 MPa H_2 and a liquid hourly space velocity (LHSV) of 2.5 h^{-1}. It was evident that the nitrogen-doped A-rGO solids were much more active than the undoped rGO, as indicated by almost complete conversion of anthracene over the former. The high activity was complemented by a much higher hydrogen transfer capability of the doped A-rGO solids, as shown in Figure 6.1.[16] Such hydrogen was determined from the amount of hydrogen

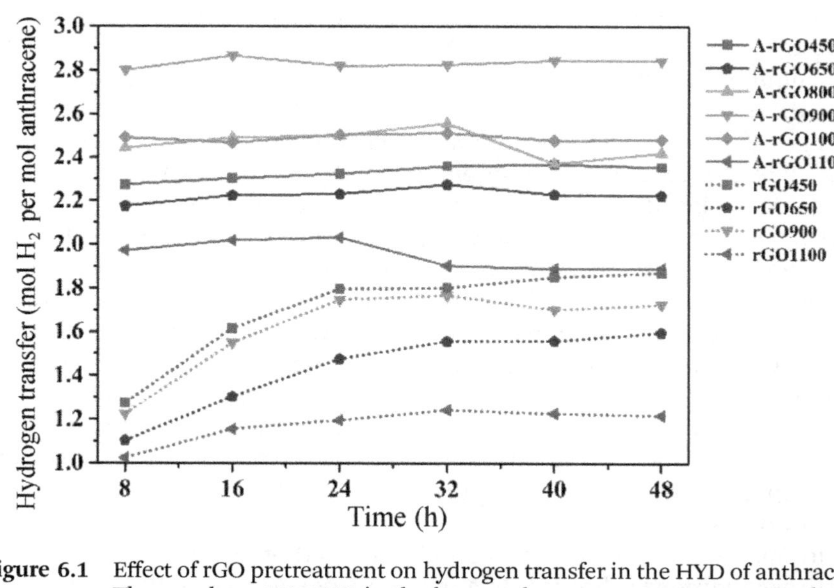

Figure 6.1 Effect of rGO pretreatment on hydrogen transfer in the HYD of anthracene. The numbers 450–1100 in the key are the treatment temperatures (°C). Reproduced from ref. 16 with permission from the Royal Society of Chemistry.

that ended up in the hydrogenated products. It was postulated that the synergy of graphitic N and the sp^2 C=C structure can enhance anthracene reactivity *via* π–π interactions whereas pyridinic N could facilitate hydrogen activation *via* dissociative adsorption. Thus, in the absence of graphitic nitrogen, the hydrogen transfer from the surface to anthracene was much less evident. In a related study, Shimoyama and Baba[18] demonstrated that the doping of r-GO with nitrogen and phosphorus significantly enhanced the absorption capability of graphene. In their study, thiophene was used as a model compound. However, catalytic HDS and/or HYD activity of the doped graphene were not determined.

6.1.2 Activity of CNTs

In line with the observations that were made for pristine graphene, a high resistance to HYD of CNTs and/or low HYD activity may be anticipated because of similarities in the basal plane surface structures. This was confirmed by Bazhenov *et al.*[19] in the HYD of single-walled carbon nanotubes (SWCNTs). However, various active sites could be created on the surface by doping pristine graphene with S, N, O and B, as confirmed by Hu *et al.*[20]

It was established that SWCNTs can reversibly absorb considerable amounts of hydrogen at moderate H_2 pressures and at room or lower temperatures.[21] Under these conditions, physisorption is the dominant mechanism of H_2 sorption.[22] However, a temperature of 450 °C and a pressure of H_2 of 90 kbar were required to load 6.8% of hydrogen while 5.4 wt% was picked up by exposing SWCNTs to H_2 at a pressure of 50 kbar at 500 °C. Most of this hydrogen was chemisorbed in the form of C–H entities.[2] In the absence of H_2, the hydrogenated SWCNTs were stable below 500 °C. However, in vacuum and above 500 °C, the H content of the SWCNTs gradually decreased. This information[6–9,19–22] confirmed that CNTs are not capable of hydrogen activation as defined for sulfided HPR catalysts.[2]

The effect of functionalization on the HYD activity of MWCNTs was investigated by Bychko *et al.*[23] The MWCNTs were prepared by CVD using ethylene and an Ni/CaO catalyst, and purified by 70% nitric acid reflux for 2 h, followed by washing and drying. The activation of MWCNTs was performed at 400 °C in a flow of H_2. This treatment resulted in the removal of carboxylic and lactone groups whereas the effect on hydroxyl and ketone groups were much less evident. Raman spectra revealed that the degree of crystallinity decreased on H_2 treatment. The catalytic activity of the H_2-treated MWCNTs was determined in the HYD of acetylene in a flow of H_2–He between 40 and 400 °C. The MWCNTs exhibited high activity and selectivity with respect to ethylene whereas ethane was a minor product. No methane formation was observed.

Information on the activity of different carbon nanomaterials under identical conditions is limited. In a relevant study by Guo *et al.*,[24] CBs, CNTs and graphene nanoplatelets (GNPs) were compared in the HDS of thiophene using tetralin as hydrogen donor. The materials were used as received without any pretreatment. Some properties of the carbon nanomaterials are given in Table 4.1. The experiments were conducted in an autoclave in the temperature range 120–180 °C. First, thiophene, tetralin and a carbon were sonicated in a reactor, which was subsequently tightly sealed, then transferred into a preheated electric oven. At the end of tests, the reactor was cooled and samples were withdrawn to determine the conversion of thiophene. The results of these tests are shown in Figure 6.2.[24] At each indicated temperature, the following activity order was maintained: CB > GNPs > CNT1 > CNT2. It should be noted that this order does not correlate with the surface areas of the carbon nanomaterials, hence other factors must be involved. For example, the small difference in activity between CB and GNPs, in spite of the large difference in surface areas, suggests a more efficient interaction of thiophene and tetralin with the GNP surface than with the CB surface. As suggested by Guo *et al.*,[24] this may involve the highly mobile π-electrons, which facilitate high electrical conductivity in accordance with the conclusions reached Wang *et al.*[25] in a study on the adsorption of thiophene on graphene in comparison with several carbon materials.

The ratio of Raman intensities in Table 4.1 indicates the extent of irregularities in the structure of the carbon nanomaterials. Thus, the D band is associated with carbon atoms close to the edges and defects whereas the G band represents the in-plane carbon atoms of a highly ordered graphene surface. Apparently, little correlation involving these parameters with catalytic activity could be established. This again confirmed that an electronic

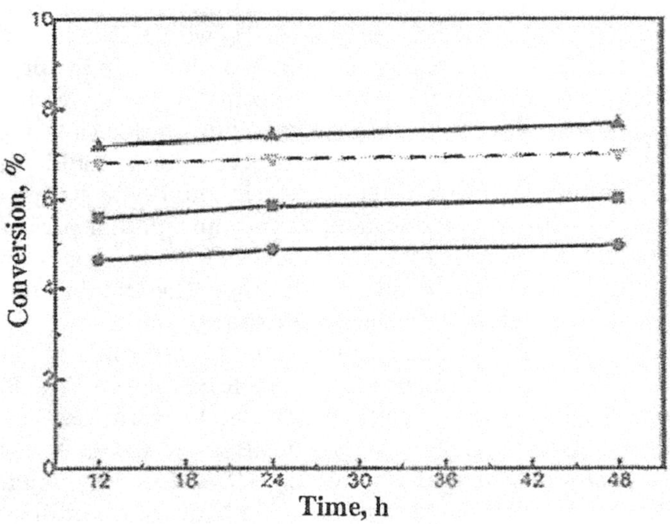

Figure 6.2 Effect of carbon nanosolids on thiophene conversion: ▲, CB; ▼, GNPs; ■, CNT1; ●, CNT2.
Reproduced from ref. 24 with permission from John Wiley and Sons, © 2017 Wiley-VCH Verlag GmbH & Co. KGaA, Weinheim.

interaction between the reactants and the surface, rather than textural properties, played a key role in catalysis.

6.1.3 Activity of Fullerenes

The absence of edge regions in fullerenes and their presence in pristine graphene, CNTs and CNFs may indicate a lower catalytic activity of the former in HYD reactions. As indicated, the exposed, not fully coordinated carbons in the edge regions are much more reactive than the carbons in fully coordinated basal planes. Thus, hydrogen activation to facilitate HYD reactions can occur more readily on graphene, CNTs and CNFs than on fullerenes. However, some experimental database is necessary to verify this assumption. Nevertheless, this situation will change dramatically once active metals are anchored on the surface of the carbon nanomaterials.

No systematic study in which fullerenes were the focus of attention as catalysts under HYD conditions was found in the literature. However, some information relevant to the activity can be obtained from work on other aspect of fullerenes. For example, in the presence of noble metals and some transition metals, C_{60} could be converted to $C_{60}H_{36}$.[26] Without active metals, methanation of C_{60} and C_{70} began at 420 °C.[27] This suggests that the capability of fullerenes alone to activate dihydrogen is low. In accordance with these observations, little catalytic activity was observed in the HCR of an extract from coal liquefaction in tetralin in the presence of dispersed fullerenes.[28] In this case, the experiments were conducted in a batch reactor at 440 °C and 19 MPa of H_2. It is believed that such conditions are unfavorable for testing fullerenes alone as catalysts.

Nishihara *et al.*[29] studied the reaction of H_2 with C_{60} between 273 and 353 K and with partial pressure varying from 0 to 120 kPa in the presence of a Pt/CB catalyst. It was confirmed that C_{60} could associate with hydrogen radicals which spilled over from the catalyst. Thus, in this case, H_2 was first dissociatively adsorbed on Pt, before hydrogen atoms could migrate on fullerenes in physical contact with Pt/CB. The amount of hydrogen adsorbed on C_{60} increased with increase in temperature and partial pressure of H_2. The strength of the reversible C_{60} interaction with hydrogen atoms was between physisorption and covalent bonding. The conditions used by Nishihara *et al.*[29] are relevant for hydrogen storage rather than for HYD or HPR. At 453 K and 5 MPa, C_{60} could be hydrogenated to $C_{60}H_{52}$ in contact with an Ru/AC catalyst.[30] Similarly, under HPR conditions (*e.g.* 673 K and 5 MPa of H_2), C_{60} mixed either with Ni/SiO_2–Al_2O_3 or with Pt/CB was gradually hydrogenated to $C_{60}H_{52}$ *via* $C_{60}H_{18}$, $C_{60}H_{36}$ and $C_{60}H_{44}$.[31,32] However, under these conditions, the stability of $C_{60}H_{44}$ and $C_{60}H_{52}$ fulleranes was low, as indicated by their fragmentation. The hydrogen bonding in $C_{60}H_x$ complexes formed at high temperatures and high H_2 pressures was covalent, with limited mobility of hydrogen atoms. This would limit the ability of $C_{60}H_x$ complexes to transfer surface hydrogen to reactant molecules to initiate HYD reactions.

6.1.4 Activity of CNFs and CNHs

The database on the catalytic activity of CNFs and CNHs in HYD reactions is limited. However, some assumptions/extrapolations can be made on the basis of available information on the activity of graphene and CNTs. It is evident from Figures 2.1–2.4 that the edge regions of CNFs account for much more of the total surface area that that of CNTs. For the latter, basal planes are the predominant contributor. Among different types of CNFs (Figure 2.4), more edge surface area is available for the platelet and spiral types compared with the ribbon-type CNFs. Both armchair and zigzag arrangements of carbon atoms in the edge regions of CNFs may be present. In the case of some irregularities, not fully coordinated carbon atoms may be exposed in the edge regions. A significantly greater reactivity of such carbon atoms compared with those present in basal planes is anticipated. This could be evident during hydrogen activation. Also, anchoring active metals could be achieved more readily in edge regions. On the basis of these assumptions, a higher HYD activity of CNFs compared with CNTs may be expected. However, it is necessary to establish a database of experimental results to confirm this assumption.

In a study by Moreira *et al.*[33] on the activity of CNFs, commercial CNFs as received and those pretreated with nitric acid were compared in the conversion of guaiacol in a batch reactor. Less than 5% overall conversion was observed at 573 K and 2 MPa of H_2. Over non-pretreated CNFs, cresols, phenol and xylenols were the major products in addition to small amounts of toluene. At 623 K, the conversion reached almost 40%, yielding the same products. Nitric acid pretreatment of CNFs resulted in an increase in selectivity for the formation of toluene and catechols.

Among carbon nanomaterials, CNHs have received the least attention. However, some information on the activity of CNHs under HYD conditions can be deduced on the basis of some similarities with CNTs. Of course, partially pretreated CNHs (*e.g.* by oxidation) may be compared with CNTs provided that both ends of particles are opened. Then, the edges of the opened CNHs, similarly to those of the opened CNTs, are much more active than basal planes. However, for both CNHs and CNTs, the contribution of the edges to the total surface area is rather small.

6.2 Nature of the Active Site of Catalysts Supported on Carbon Nanosupports

The type of interaction of active metals with carbon nanosupports defines the geometry of the active sites. It has been generally accepted that the basal planes of pristine nanocarbons such as graphene, CNTs, CNFs and CNHs are chemically inactive because of the high stability of the π-conjugated system present. On the other hand, the edges of the nanocarbons possess some activity for adsorption of various reactants and metals and as such they may be active in some catalytic reactions. Both the adsorption capability and catalyst activity may be enhanced by creating defects, *e.g.* by doping. For example, doping with N modifies the surface of carbon nanomaterials by creating a graphitic and a pyridinic nitrogen, as depicted in Figure 6.3. According to Nakamura and Kondo,[34] the carbons in the vicinity of graphitic and pyridinic nitrogen are acidic and basic sites, respectively. According to Liu *et al.*,[16] in the absence of catalytically active metals, graphitic nitrogen facilitates π–π interactions with aromatic rings (*e.g.* anthracene) adsorbed on N-doped graphene whereas pyridinic nitrogen favors the dissociation of H_2 to active surface hydrogen. Therefore, doping with nitrogen has a dramatic effect on the electronic properties of the originally inert surface of carbon nanomaterials.

As was pointed out earlier, the availability of other dopants (*e.g.* S, O, B, P) is the basis for modifying the electronic properties of the surface of carbon nanomaterials with the aim of enhancing the interaction with catalytically active metals. For example, the surface of CNTs consisting of a π-conjugated system can interact with catalytically active metals. Nakamura and Kondo[34] showed that in the case of Pt and Pt–Ru alloy, this interaction involves a π state of carbon and a d state of the metals. The π–d hybridization modifies the electronic structure of Pt. Because the π-conjugated local structure of the surface can be modified (*e.g.* by doping), active metals can attain different electronic states. Consequently, the same active metal can exhibit different activities and selectivities.

In the case of CNTs, the interaction with active metals is influenced by the diameter of the CNTs. Therefore, because the size of nanoparticles of active metals anchored on carbon nanosupports may vary from 1 to 7 nm, little interaction between the active metals and the support can be established on SWCNTs having a diameter of 1 nm or less. According to Nakamura and Kondo,[34] such nanoparticles of metals can be homogeneously dispersed on CNTs having diameters between 10 and 30 nm. For CNT diameters >100 nm, the effect of the support on metal nanoparticles is absent. Pt particles with diameter <1 nm could be readily deposited on the basal plane of graphene.[35,36]

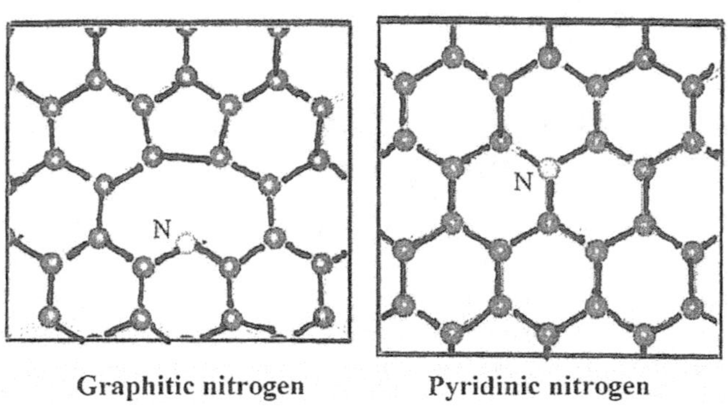

Figure 6.3 Structures of N-doped graphene.

The interaction of active metals with the carbons located in peripheral parts of graphene and also at the edges and/or ends of CNTs and CNHs differs from that with the carbon in basal planes. Not fully coordinated carbons in the former locations can coordinate with active metals in a similar manner to CB, *i.e. via* weak coulombic interactions compared with π–d interactions in basal planes.[34] Therefore, the nature of these active sites must differ from that of the sites located in basal planes. Apparently, less hindered access of reactant molecules to the active sites at the ends/edges of CNTs may be anticipated. However, because the contribution of the surface area of the ends of CNTs to the overall surface area is low, the active sites located there may not play an important role in overall catalysis, although by decreasing the length of the CNTs the contribution of the surface area of the ends to the total surface area can be increased.

The curvature of the surfaces of CNTs, CNFs and CNHs compared with the relatively flat surface of graphene may also influence the geometry of active sites. The effect of the surface curvature becomes more evident for the active sites located on the exterior compared with those inside these support particles. Whereas the π–d interaction of active metals with carbons can be retained, the accessibility of reactant molecules to active sites in the interior of CNTs, CNHs and CNFs depends on the diameter of these support particles. In this case, not only the effect of reactant size but also the effect of the opposing curvature compared with that on the exterior of the carbon nanosupports play a role in defining the geometry of active sites.

The structure of carbon nanosupports determines the geometry of the active sites. This may be illustrated using the structure of CNFs shown in Figure 2.4. For example, the ribbon-type CNFs consist of two large basal plane regions together with two long and two short edge regions with one edge possessing an armchair and the other a zigzag configuration. Apparently, the long edge region will adopt a zigzag orientation. The platelet CNFs consist of two edges of similar length oriented equally in either the armchair or zigzag configuration. The spiral-like CNFs consist of platelets oriented similarly as in the ribbon-like CNFs. The geometric form of these CNFs causes a strain, leading to a change in spacing between platelets. This changed spacing modified the interaction with active metals.[37] The orientation of platelets in the CNF structures (Figure 2.4) had different influences on Ni particles in Ni/CNF catalysts. This was evident from the different *trans*- to *cis*-2-butene ratios observed in the HYD of 1-butene, *i.e.* the following decreasing order was observed: ribbon-like > spiral > platelet CNFs. The activity and selectivity trends changed dramatically when 1,3-butadiene was used as reactant. In this case, Ni supported on ribbon CNFs was the least active. These observations indicate some complexity in the nature of the active sites even for such simple reactions as the HYD of butene and butadiene.

A unique effect of CNFs on Ni metal compared with that of AC and γ-Al_2O_3 supports was reported by Baker and co-workers.,[38,39] who prepared Ni(5%) catalysts supported on these supports by incipient wetness impregnation.

The activity and selectivity of these catalysts in the HYD of 1-ethene and 1,3-butadiene (atmospheric pressure H_2, 80 °C, continuous) were compared. In this case, the CNFs consisted of graphene platelets stacked in a perpendicular direction with respect to the axis of the fibers, *i.e.* platelet-type CNFs in Figure 2.4. With such an arrangement, only peripheral parts of the graphene layers consisting of edges were available to interact with active metals during catalyst preparation. These edges accounted for a large proportion of the CNF surface. Spectroscopic evaluation of the Ni/CNF surface revealed an even distribution of Ni with a well-defined hexagonal shape of the particles. For an Ni/AC catalyst, a high distribution was also observed, but the size of the Ni particles was much smaller. Moreover, the morphology of the Ni particles was mixed, without any preferred geometric form. A high distribution and very small size of Ni particles, *i.e.* five times smaller than those on CNFs, were observed on γ-Al_2O_3. Because of the very small size, the morphological characteristics of Ni particles on γ-Al_2O_3 could not be described. In the HYD of 1-butene and 1,3-butadiene, the Ni/CNF catalyst exhibited the best performance and Ni/Al_2O_3 the worst. This is in contradiction with the Ni particle distribution being the highest on Ni/Al_2O_3 and the lowest on Ni/CNF catalysts. Hence, rather than particle size, the morphology of Ni particles induced by the supports played a dominant role in defining the activity and selectivity of the Ni catalysts. Apparently, the catalysis of HYD reactions over Ni catalysts can be influenced by electronic effects induced by the supports on Ni metal. In this regard, the beneficial effect facilitated by CNFs was much more evident than that by γ-Al_2O_3 and AC.

One of the functions of the active sites in HYD reactions is the transformation of H_2 into active surface hydrogen entities (*e.g.* Me–H and C–H; Me = active metal). In this regard, a higher efficiency of noble metals as active catalysts compared with transition metals has been observed. Moreover, the hydrogen activation is favored on carbon nanosupports to a greater extent than on oxidic supports.[40] This is confirmed by the results presented in Figure 6.4, showing the amount of H_2 desorbed in the TPD of catalysts previously exposed to H_2. Quantification of these amounts revealed that the H_2 adsorption on an Ru/CNT catalyst was twice that on an Ru/Al_2O_3 catalyst, and on the other Ru catalysts the amounts of H_2 adsorbed were very small.

Intimate contact of reactants with active metals is of particular importance because of the presence of active hydrogen. Thus, most of the active hydrogen is on and/or in the proximity of active metals. The type of transition state formed between the reactant and the active site determines the structure of products. Therefore, the structure of the reactant plays an important role in defining the geometry of the active site. The adsorption of α,β-unsaturated aldehydes on active metals is of particular interest because of their frequent use as model compounds. According to Delbecq and Sautet,[41] adsorption on an active site may occur *via* coordination with the active metal involving either C=C or C=O double bonds. In this case, both atoms of the C=C and C=O bonds are attached to the active metal.

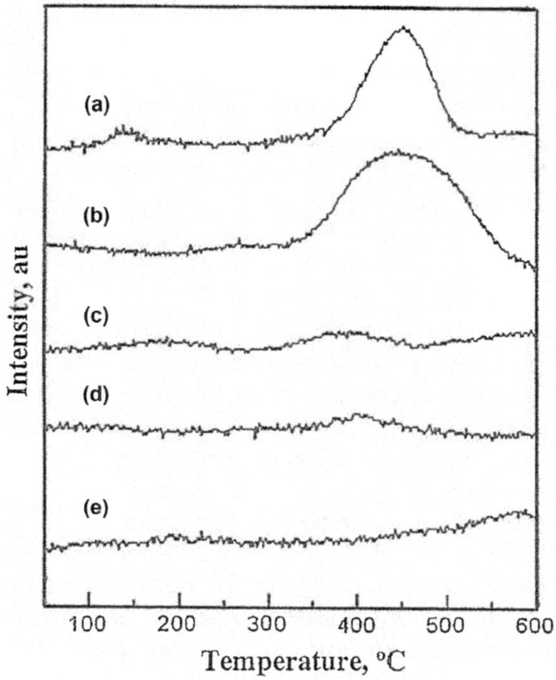

Figure 6.4 H_2 TPD profiles of Ru catalysts: (a) Ru/Al_2O_3; (b) Ru/CNT; (c) Ru/MgO; (d) Ru/CeO_2; (e) Ru/SiO_2.
Reproduced from ref. 40 with permission from Springer Nature, Copyright 2009.

In addition, the top adsorption involving the oxygen atom of the aldehyde group directly attached to Me, *i.e.*

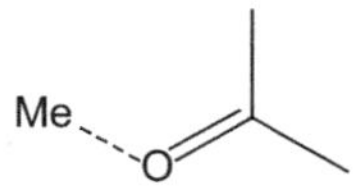

may be present. On the basis of theoretical calculations by Delbecq and Sautet,[42] substitution by methyl or phenyl such as in acrolein, giving crotonaldehyde and CALD, respectively, would result in a significant decrease in the binding energy involving the C=C double bond. At the same time, an adverse effect on the bonding energy involving the C=O bond would be much less evident. The substitution would have little effect on the binding energy of the top adsorption. For Pt-containing catalysts, adsorption *via* the C=O double bond relative to the C=C double bond could be enhanced by alloying active metals, as was observed for Pt metal and Pt–Fe alloy.[42] In the case of the HYD of CALD, this would enhance the formation of unsaturated alcohols such as CALC (Figure 5.2).

Geometrically, an active site may comprise both an active metal and a bare surface in the proximity of the active metal. For catalysts supported on carbon nanosupports, active metals play a decisive role during the adsorption and activation of hydrogen. To various extents, reactant molecules can be adsorbed on both the active metal and bare surface to form a transition state from which the product molecules arise after being desorbed from the surface. The type of products arising from the activated adsorption of reactants depends on the orientation of the reactants towards the active site. The experimental results reported by de Jong[43] and Toebes *et al.*[44] are used to support this concept of active sites. They used the HYD of CALD (batch reaction, 383 K, 4.8 MPa of H_2) to investigate the effect of the catalyst surface structure on conversion and selectivity. For this purpose, they prepared a Pt(3.6%)/CNF catalyst by the deposition–precipitation method.[45] The catalyst was treated at different temperatures under N_2. Figure 6.5[43] shows an increase in the overall conversion of CALD with increase in the pretreatment temperature. Obviously, this change in catalyst activity must be associated with the change in geometry of the active site. First, the concentration of O-containing groups on CNFs was decreased significantly on thermal pretreatment. Second, an increase in the size of the Pt nanoparticles from ~1 to ~2 nm was noted. These changes influenced the distribution of products in the HYD of CALD (Figure 5.2) over non-pretreated Pt/CNT compared with that over Pt/CNT treated at 973 K. Thus, over the latter catalyst, with an initial concentration range of CALD of 103–345 mol m^{-3} at 50% conversion, the selectivity to CALC approached about 10% compared with about 40% over the untreated Pt/CNF catalyst. At the same time, the selectivity with respect to HCALD over the thermally treated and untreated

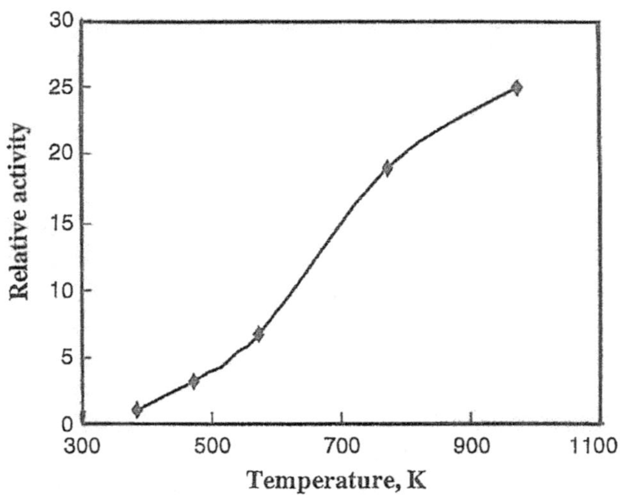

Figure 6.5 Effect of pretreatment temperature on activity of Pt/CNF in the HYD of CALD.
Reproduced from ref. 24 with permission from John Wiley and Sons, © 2017 Wiley-VCH Verlag GmbH & Co. KGaA, Weinheim.

catalysts approached 80 and 40%, respectively. According to de Jong,[43] with the untreated Pt/CNF, the adsorption of CALD *via* the benzene ring on the catalyst surface was partially impaired by the presence of O-containing groups. Once these groups had been removed, the adsorption of the benzene ring in CALD on the surface became stronger. This favored the adsorption of CALD *via* the C=C bond relative to the C=O bond on Pt metal. On the other hand, the smaller size of Pt nanoparticles on the untreated Pt/CNF catalyst favored adsorption *via* the C=O bond, leading to the formation of CALC.

Studies involving unsupported active phases, which were also investigated as part of the catalysts supported on carbon nanosupports, may contribute to the understanding of active sites. Of particular interest is the transition state consisting of extensively used oxygenates. In this case, a difference in the activated interaction between the C–O and C–C bonds may be expected. This will influence the initial rate of the reactant transformation. This was demonstrated in several studies. For example, using the DFT method, Ren *et al.*[46] estimated the activation barriers for C=O and C–C bonds on close-packed [0001] Mo_2C to be 0.61 and 1.21 eV, respectively. This suggests that for molecules such as aldehydes and ketones, the C=O bond would be cleaved preferentially compared with the adjacent C–C bond. This was confirmed experimentally in the HDO of propanal over a thin Mo_2C film by the formation of propylene rather than propane even in the absence of H_2. The eliminated oxygen remained attached to the active phase. Consequently, the catalyst activity declined with time on-stream. As soon as H_2 was introduced, a steady catalyst performance was observed, confirming the active site renewal due to oxygen removal from the active site. This was confirmed using 1-propanol, 2-propanol and acetone at 573 K in a flow of H_2 by the formation of propylene as the main product with only traces of C_2 hydrocarbons being formed. Because of some electronic similarities between Mo_2C and Pt,[47] a similar interaction of the C=O bond with Pt may be anticipated. The preference of C=O bond cleavage relative to C–C bond cleavage was also confirmed in the HDO of furfural using active phases such as Fe/Pt(111) and Fe/Mo_2C/Mo(110).[48] A strong interaction on both surfaces led to enlargement and/or weakening of the C=O bond and eventually cracking to yield methylfuran. At the same time, the adjacent C–C bond was shortened.

6.3 Methods for Catalyst Preparation

The importance of the functionalization and doping of carbon nanosupports with regard to the performance of catalysts is again emphasized. This can be illustrated using the results of Plomp *et al.*,[49] who showed that by introducing O-containing groups on CNFs, the size of Pt and Pd particles could be controlled. For example, in the presence of O-containing groups, larger sized Pt and Pd nanoparticles could be deposited on the catalysts, *i.e.* 3.5 nm. Once the O-containing groups had been removed by thermal treatment, the size of the particles decreased to 2.0 nm. Consequently, the catalyst activity and selectivity in the HYD of CALD to CALC (Figure 5.2) increased. This suggests that a delicate balance between functionalization and subsequent thermal treatment must be established in order to attain optimal catalyst performance.

Brief accounts of the preparation, properties, functionalization and doping of carbon nanomaterials used as catalyst supports were given above. In the following, the focus is on the methodology for the deposition of active metals on carbon supports. The most widely used technique to achieve the deposition of metals on supports has been incipient wetness impregnation. This method has also been widely used for the preparation of carbon nanomaterial-supported catalysts. In this case, after being pretreated (*e.g.* by nitric acid purification/oxidation), the support is impregnated with an aqueous solution of metal precursor. This is usually followed by drying, calcining and reduction. Peroxidation with nitric acid resulted in an improvement of the hydrophilic nature of the carbon surface. It also ensured that active metals are anchored on the carbon surface. A strong interaction of active metals with carbon supports must be maintained to ensure the stability of the catalyst during reaction.

In the following, several examples of methodologies that have been used for the deposition of active metals on carbon nanosupports are presented. The methods of catalyst preparation used in the reviewed studies are also briefly described. It is evident that similar methodologies have been used for catalyst preparation regardless of the type of carbon nanosupport.

It was indicated above that the size of nanoparticles of active metals is among the parameters that influence the course of catalytic reactions. Here the conditions applied during the catalyst preparation can play a decisive role. This was demonstrated in a study on the preparation of Pd/CNT catalysts by Lamme *et al.*,[50] who investigated the effects of different methods of Pd deposition on CNTs and subsequent activation. First, $(NH_3)_4Pd(NO_3)_2$, used as the Pd precursor, was deposited from aqueous solution onto nitric acid-treated CNTs using different methods such as incipient wetness impregnation, deposition precipitation and ion adsorption. The aim was to deposit 5.0–6.0 wt% of Pd. The solids were activated by thermal treatment in N_2 at 500 °C and also by liquid-phase reduction using either formaldehyde or sodium borohydride. The activation in N_2 resulted in the formation of homogeneously distributed Pd particles with an average size of 4 nm. The liquid-phase reduction in formaldehyde yielded particles of ~1 nm in size whereas the reduction with sodium borohydride produced a catalyst with a bimodal size distribution, *i.e.* ~1 nm and >10 nm. The performance of these catalysts was evaluated in the HYD of CALD.

Kvande *et al.*[51] compared ion exchange, homogeneous deposition–precipitation and impregnation as methods for the preparation of Pt catalysts supported on fishbone CNFs. The small amount of Pt that could be deposited was a limitation of these methods. On the other hand, up to 24% of Pt with a particle size of 2–3 nm could be deposited using colloidal methods. The deposition–precipitation method was used to deposit Ni nanoparticles on CNFs.[52] In this case, Ni particles with an average size of 9 nm were obtained.

Zhang *et al.*[53] synthesized CNF-supported Pt catalysts using a homogeneous deposition–precipitation (HDP) method. An acidified suspension of CNFs in water was maintained at 90 °C and mixed vigorously with urea and a Pt precursor such as $Pt(NH_3)_4(NO_3)_2$ for 18 h. The Pt-loaded CNFs were filtered, washed with demineralized water, dried at 120 °C and reduced with H_2 at 200 °C for 1 h. In a study by Jung *et al.*,[54] Pt/CNF catalysts were prepared by four different methods, *i.e.* conventional wetness impregnation, HDP, ion adsorption and colloidal microwave methods. The activity and selectivity of the Pt/CNFs catalysts prepared by these methods were evaluated in the HYD of CALD (Figure 5.2). As the results in Table 6.1[54] show, the most active Pt/CNF catalyst was obtained by the colloidal microwave method. In this case, the selectivity with respect to CALC was determined at 80% conversion. In a study by Toebes *et al.*,[55] the HDP method was compared with the ion adsorption method in the preparation of Pt and Ru catalysts supported on CNFs.

In a comprehensive study on the preparation of CNT-supported catalysts, Nguyen and Serp[56] evaluated the effects of different precursors (*e.g.* nitrate, chloride and organometallic complexes) for active metals such as Ru, Pd and Co and also the type of solvents on the filling of metal nanoparticles in the channels of CNTs. The starting material (CNT0) had od 10.9 and id 5.5 nm. CNT0 was treated with nitric acid at 140 °C to obtain CNT1 and also ball-milled to obtain CNT2. Both CNT1 and CNT2 were thermally treated in Ar at 900 °C for 2 h to obtain the corresponding CNTHTs. These supports were mixed with metal precursors, ultrasonicated and stirred at room temperature for 12 h. After drying at 110 °C, the samples were reduced in a flow of H_2–Ar at 300 °C for 2 h. The filling yields (see Table 6.2)[56] were estimated from conventional 2D transmission electron micrographs on the basis of the number of particles in the channels and the total number of particles. The best filling yields were obtained for nitric acid-treated nanotubes (CNT1). It was pointed out that the confinement of metal particles was governed by the openings of the CNTs and the affinity of the metal precursors during impregnation. Hence, to favor confinement, the affinity to the inner surface of the CNT channels should be greater than that to the outer surface. In addition, the type of solvent in terms of surface tension and viscosity plays an

Table 6.1 Methods for the deposition of Pt on CNF supports.[54]

Method	Pt content/wt%	Activity/mol $(kg\,Pt)^{-1}\,s^{-1}$	Selectivity/%
Wetness impregnation	5.6	0.14	18
Homogeneous deposition–precipitation	5.7	0.19	11
Ion adsorption	5.2	0.15	10
Colloidal microwave processing	2.2	0.95	70

Table 6.2 Metal (Ru, Co and Pd) loading, nanoparticle size and filling yields for CNTs.[56]

	Metal loading/%			Nanoparticle size/nm			Filling yield/%		
Support	Ru	Co	Pd	Ru	Co	Pd	Ru	Co	Pd
CNT0	2.17	3.17	3.25	2.21	3.48	4.15	19	15	13
CNT1	0.99	2.72	1.24	1.02	2.72	1.15	73	64	72
CNT2	2.01	3.28	3.08	1.87	3.25	4.01	48	38	35
CNT1HT	1.41	2.94	1.64	1.25	2.88	2.48	60	53	58
CNT2HT	1.23	3.02	2.48	1.47	2.95	3.85	66	62	60

important role. The results in Table 6.2[56] are for impregnation in acetone using metal precursors such as $[Ru(NO)(NO_3)_3]$, $Co(NO_3)_2{\cdot}6H_2O$ and $Pd(NO_3)_2(H_2O)_2$ in acetone.

Giordano *et al.*[57] described a method for carboxylate group-mediated grafting of Rh metal on MWCNTs. As part of this procedure, the support was treated with nitric acid to introduce carboxyl groups. A suspension of the acid-functionalized MWCNTs in toluene was mixed with an aqueous solution of sodium carbonate. After sonication at 313 K for 2 h, the solid was filtered, washed and dried at 383 K to obtain MWCNT-COONa. This solid was mixed with $[Rh_2Cl_2(CO)_4]$ precursor in toluene and stirred at 333 K for 3 h. The Rh-grafted solid was obtained after washing with *n*-hexane and vacuum drying. This solid was reduced in N_2–H_2 at 573 K for 3 h. As the final step, the solid was washed to remove the NaCl that was formed during grafting. Spectroscopic evaluation of the solid revealed no evidence of carbonyl groups. Moreover, grafting ensured a high dispersion of Rh particles on the surface. In contrast, large aggregates of Rh particles were present on Rh/MWCNT prepared without grafting. The catalytic activity of this catalyst in the HYD of CALD was negligible compared with the high activity of Rh/MWCNT prepared by grafting.

Ni *et al.*[58] used a microwave-assisted thermolytic method for the preparation of Ru catalysts supported on pristine CNTs and functionalized CNTs (CNT-f). In this case, the CNTs and an Ru precursor such as $Ru_3(CO)_{12}$ were mechanically mixed in an agate mortar for 20 min. The mixture was transferred into a quartz reactor and fluidized with Ar at room temperature for 2 h. The reactor was then placed in a microwave oven operating at 2.45 GHz with a power of 800 W. Using this method, variable amounts of Ru could be deposited on the CNTs and CNT-f by varying the amount of precursor in the mixture with CNTs. On the nonfunctionalized CNT support, Ru particles with a size range of 2–4 nm could be readily deposited. Under similar conditions, the size of Ru particles deposited on the functionalized CNTs ranged from 3 to 6 nm.

Attempts have been made to develop hybrid catalysts by immobilizing organometallic complexes inside CNTs. In this case, the proportion of CNTs with both ends opened is an important factor, in addition to a suitable diameter ensuring easy diffusion of the complexes into the interior of the CNTs. For example, the Pd–tridecylamine complex was successfully confined in CNTs, which subsequently exhibited significantly enhanced activity for the HYD of cyclohexene.[59] In a similar study, the $[Rh(COD)NH_2CH_2CH_2NH(CH_2)_3Si(OCH_3)_3]^+BF_4^-$ complex was confined in CNTs and also used for the HYD of cyclohexene.[60] The confinement of the complex also resulted in an improvement in HYD activity.

Recently, Hajjar *et al.*[61] gave detailed accounts of the preparation of several Co–Mo catalysts supported on graphene oxide (GO), which was prepared by the modified Hummers method. In this case, graphite was treated with a mixture of concentrated HNO_3 and H_2SO_4 in an ice bath and further treated with $NaNO_3$ and $KMnO_4$. This resulted in expansion of the graphite layers. After the addition of water, the mixture was oxidized with H_2O_2 to produce GO, which was separated by exfoliation with the aid of ultrasonic vibration. The cake of GO was filtered, washed with deionized water and dried before being used for catalyst preparation. The active metals were added using an incipient wetness impregnation method and a hydrothermal method using $NH_4(H_2O)_6Mo_7O_{24}$ and $Co(NO_3)_2$ as precursors. After impregnation, the solids were dried and calcined at 300 °C in N_2. The hydrothermal preparation was carried out in an autoclave. In this case, a suspension of GO and metal precursors was mixed vigorously with citric acid and maintained at 180 °C for 16 h. The solid obtained was dried and calcined as above. Furthermore, a D-optimal experimental design involving three numerical and categorical factors was undertaken to identify the most active catalyst. Such a catalyst contained 10 wt% of metal with a Co:Mo ratio of 0.3. In another method used by the same group,[62] a vigorously stirred suspension of GO with a solution of precursors was sprayed into a pyrolyzer preheated to 950–1000 °C in a flow of N_2. The solid product was collected in water, filtered and dried. The activity of these catalysts, determined in the HDS of naphtha, is discussed in Chapter 9, Section 9.2.4.

Xu *et al.*[63] used a hexamethonium-assisted hydrothermal method to transform the negatively charged surface of GO into a positively charged surface. This improved the interaction of active metals such as Mo and Co with the surface. The concentration of positive charges could be adjusted using different amounts of hexamethonium. The modified GO was used as the support for the preparation of a CoMoS catalyst by impregnation. During the final steps, the Co–Mo/GO was vacuum dried at 60 °C, heated under N_2 at 450 °C, then sulfided using a CS_2–heptane mixture at 360 °C and 2 MPa.

References

1. O. Mohammadi, M. Golestanzadeh and M. Abdouss, *New J. Chem.*, 2017, **41**, 11471–11493.
2. M. Breysse, E. Furimsky, S. Kasztelan, M. Lacroix and G. Perot, *Cat. Rev.: Sci. Eng.*, 2002, **44**, 651–735.
3. F. H. Yang, A. J. Lachawiec and R. T. Yang, *J. Phys. Chem. B*, 2006, **110**, 6236–6244.

4. M. Kayanuma, U. Nagashima, H. Nishihara, T. Kyotani and H. Ogawa, *Chem. Phys. Lett.*, 2010, **495**, 251–255.
5. D. C. Elias, R. R. Nair, T. M. G. Mohiuddin, S. V. Morozov, P. Blake, M. P. Halsall, A. C. Ferrari, D. W. Boukhvalov, M. I. Katsnelson, A. K. Geim and K. S. Novoselov, *Science*, 2009, **323**, 610–613.
6. L. Kyhl, R. Bisson, R. Balog, M. N. Groves, E. L. Kolsbjerg, A. M. Cassidy, J. H. Jørgensen, S. Halkjær, J. A. Miwa, A. Grubišić Čabo, T. Angot, P. Hofmann, M. A. Arman, S. Urpelainen, P. Lacovig, L. Bignardi, H. Bluhm, J. Knudsen, B. Hammer and L. Hornekaera, *ACS Nano*, 2018, **12**, 513–520.
7. Z. Ao and F. Peeters, *J. Phys. Chem. C*, 2010, **114**, 14503–14509.
8. D. Smith, R. T. Howie, I. F. Crowe, C. L. Simionescu, C. Muryn, V. Vishnyakov, K. S. Novoselov, Y.-J. Kim, M. P. Halsall and E. Gregoryanz, *ACS Nano*, 2015, **9**, 8279–8283.
9. V. B. Parambhath, R. Nagar, K. Sethupathi and S. Ramaprabhu, *J. Phys. Chem. C*, 2011, **115**, 15679–15685.
10. A. Primo, A. Franconetti, M. Magureanu, N. B. Mandache, C. Bucur, C. Rizescu, B. Cojocaru, V. I. Parvulescu and H. Garcia, *Green Chem.*, 2018, **20**, 2611–2623.
11. M. Shaikh, S. K. Singh, S. Khilari, M. Sahu and K. V. S. Ranganath, *Catal. Commun.*, 2017, **106**, 64–67.
12. E. Furimsky, *Ind. Eng. Chem. Res.*, 2017, **56**, 11359–11371.
13. E. Furimsky, *Ind. Eng. Chem. Res.*, 2013, **52**, 17695–17713.
14. E. G. Mission, A. T. Quitain, Y. Hirano, M. Sasaki, M. J. Cocero and T. Kida, *Catal. Sci. Technol.*, 2018, **8**, 5434–5444.
15. D. Wang, W. Niu, M. Tan, M. Wu, X. Zheng, Y. Li and N. Tsubaki, *ChemSusChem*, 2014, 7, 1398–1406.
16. R. Liu, F. Li, C. Chen, Q. Song, N. Zhao and F. Xiao, *Catal. Sci. Technol.*, 2017, 7, 1217.
17. W. S. Hummers and R. E. Offeman, *J. Am. Chem. Soc.*, 1958, **80**, 1339–1342.
18. I. Shimoyama and Y. Baba, *Carbon*, 2016, **98**, 115–125.
19. A. V. Bazhenov, T. N. Fursova, I. O. Bashkin, V. E. Antonov, I. V. Kondrat'eva, A. V. Krestinin and Y. M. Shul'ga, *Fullerenes, Nanotubes, Carbon Nanostruct.*, 2006, **14**, 165–170.
20. H. Hu, J. H. Xin, H. Hu, X. Wang and Y. Kong, *Appl. Catal.*, 2015, **492**, 1–9.
21. M. J. Heben and A. C. Dillon, *Appl. Phys. A: Mater. Sci. Process.*, 2001, **72**, 133–142.
22. B. K. Pradhan, G. U. Sumanasekera, K. W. Adu, H. E. Romero, K. A. Williams and P. C. Eklund, *Physica B*, 2002, **323**, 115–121.
23. I. Bychko, A. Abakumov, N. Lemesh and P. Strizhak, *ChemCatChem*, 2017, **9**, 4470–4474.
24. K. Guo, M. Gu and Z. Yu, *Energy Technol.*, 2017, **5**, 1228–1234.
25. L. F. Wang, R. T. Yang and C. L. Sun, *AIChE J.*, 2013, **59**, 29–32.
26. T. Osaki, T. Tanaka and Y. Tai, *React. Kinet. Catal. Lett.*, 2003, **79**, 319–324.

27. T. Osaki, T. Tanaka and Y. Tai, *React. Kinet. Catal. Lett.*, 2003, **78**, 217–223.
28. S.-F. Zhang, J. R. Shearman, M. Domin, M.-J. Lazaro and R. Kandiyoti, *Fuel*, 1997, **76**, 207–214.
29. H. Nishihara, T. Simura and T. Kyotani, *Chem. Commun.*, 2018, **4**, 3327–3330.
30. K. Shigematsu, K. Abe, M. Mitani and K. Tanaka, *Fullerenes Sci. Technol.*, 1993, **1**, 309–318.
31. S. M. Luzan and A. V. Talyzin, *Fullerenes Sci. Technol.*, 2012, **20**, 319–323.
32. Y. Sui, J. Qian, J. Zhang, X. Zhou, Z. Gu, Y. Wu, H. Fu and J. Wang, *Fullerenes Sci. Technol.*, 1996, **4**, 813–818.
33. R. Moreira, E. Ochoa, J. L. Pinilla, A. Portugal and I. Suelves, *Catalysts*, 2018, **8**, 127–141.
34. J. Nakamura and T. Kondo, *Top. Catal.*, 2013, **56**, 1560–1568.
35. R. Siburian, T. Kondo and J. Nakamura, *J. Phys. Chem. C*, 2013, **117**, 3635–3645.
36. E. J. Yoo, T. Okada, T. Akita, M. Kohyama, J. Nakamura and I. Honma, *Nano Lett.*, 2009, **9**, 2255–2259.
37. N. M. Rodriguez, A. Chambers and R. T. K. Baker, *Langmuir*, 1995, **11**, 3862–3866.
38. A. Chambers, T. Nemes, N. M. Rodriguez and R. T. K. Baker, *J. Phys. Chem. B*, 1998, **102**, 2251–2258.
39. N. M. Rodriguez, M. S. Kim and R. T. K. Baker, *J. Phys. Chem.*, 1994, **98**, 13108–13111.
40. W. Deng, X. Tan, W. Fang, Q. Zhang and Y. Wang, *Catal. Lett.*, 2009, **133**, 167–174.
41. F. Delbecq and P. Sautet, *J. Catal.*, 1995, **152**, 217–236.
42. F. Delbecq and P. Sautet, *J. Catal.*, 1996, **164**, 152–165.
43. K. P. de Jong, *Oil Gas Sci. Technol.*, 2006, **61**, 527–534.
44. M. L. Toebes, T. A. Nijhuis, J. Hájek, J. H. Bitter, J. A. van Dillen, D. Y. Murzin and K. P. de Jong, *Chem. Eng. Sci.*, 2005, **60**, 5682–5695.
45. M. K. van der Lee, A. J. van Dillen, J. H. Bitter and K. P. De Jong, *J. Am. Chem. Soc.*, 2005, **127**, 13573–13582.
46. H. Ren, W. Yu, M. Salciccioli, Y. Chen, Y. Huang, K. Xiong, D. G. Vlachos and J. G. Chen, *ChemSusChem*, 2013, **6**, 798–801.
47. J. R. Kitchin, J. K. Nørskov, M. A. Barteau and J. G. Chen, *Catal. Today*, 2005, **105**, 66.
48. W. Wan, Z. Jiang and J. G. Chen, *Top. Catal.*, 2018, **61**, 439–445.
49. A. J. Plomp, H. Vuori, A. O. I. Krause, K. P. de Jong and J. H. Bitter, *Appl. Catal.*, 2008, **351**, 9–15.
50. W. S. Lamme, J. Zečević and K. P. de Jong, *ChemCatChem*, 2018, **18**, 1552–1555.
51. I. Kvande, S. T. Briskeby, M. Tsypkin, M. Rønning, S. Sunde, R. Tunold and D. Chen, *Top. Catal.*, 2007, **1–4**, 45.
52. J. H. Bitter, M. K. van der Lee, A. G. T. Slotboom, A. J. van Dillen and K. P. de Jong, *Catal. Lett.*, 2003, **89**, 139–142.

53. Y. Zhang, M. L. Toebes, A. vanderEerden, W. E. O'Grady, K. P. de Jong and D. C. Koningsberger, *J. Phys. Chem.*, 2004, **108**, 18509–18519.
54. A. Jung, A. Jess, T. Schubert and W. Schutz, *Appl. Catal.*, 2009, **362**, 95–105.
55. M. L. Toebes, M. K. van der Lee, L. M. Tang, M. H. Huis in 't Veld, J. H. Bitter, A. J. van Dillen and K. P. de Jong, *J. Phys. Chem. B*, 2004, **108**, 11611–11619.
56. T. T. Nguyen and P. Serp, *ChemCatChem*, 2013, **5**, 3595–3603.
57. R. Giordano, P. Serp, P. Kalck, Y. Kihn, J. Schreiber, C. Marhic and J.-L. Duvail, *Eur. J. Inorg. Chem.*, 2003, **3**, 610–617.
58. X. Ni, B. Zhang, C. Li, M. Pang, D. Su, C. T. Williams and C. Liang, *Catal. Commun.*, 2012, **24**, 65–69.
59. I. Such-Basáñez, M. C. Román-Martínez and P. Serp, *Microporous Mesoporous Mater.*, 2016, **225**, 378–384.
60. L. J. Lemus-Yegres, M. C. Román-Martínez, I. Such-Basáñez and C. Salinas-Martínez de Lecea, *Microporous Mesoporous Mater.*, 2008, **109**, 305–316.
61. Z. Hajjar, M. Kazemeini, A. Rashidi and S. Soltanali, *J. Taiwan Inst. Chem. Eng.*, 2017, **78**, 566–575.
62. Z. Hajjar, M. Kazemeini, A. Rashidi, S. Soltanali and F. Bahadoran, *J. Anal. Appl. Pyrolysis*, 2017, **123**, 144–151.
63. J. Xu, Y. Guo, T. Huang and Y. Fan, *Appl. Catal., B*, 2019, **244**, 385–395.

CHAPTER 7

Catalysts Supported on Carbon Nanotubes

The database on the preparation and testing of CNT-supported catalysts is much more extensive than that of catalysts supported on CNFs, CNHs, graphene and fullerenes. The advantages of CNT supports compared with other supports, as part of HYD catalysts, were discussed by Shuit *et al.*[1] Both SWCNTs and MWCNTs have been evaluated as supports for the preparation of catalysts containing noble metals, transition metals and rare earth metals. Testing has been conducted under conditions typical of those encountered in HYD for the production of fine chemicals and also in HPR to produce fuels. Thus, both batch and continuous systems have been used. Polar and nonpolar organic solvents and water have been used as reaction media.

7.1 Noble Metals Supported on CNTs

In comparison with other metals supported on carbon nanomaterials, catalysts comprising noble metals supported on CNTs have been evaluated the most extensively. In most studies, oxygenates typical of those derived from various types of biomass have been used as reactants. In this case, products derived from the components of lignocellulosic biomass, *i.e.* lignin, cellulose and hemicellulose, in addition to those of vegetable oil and algae biomass origin, have been used as model reactants. To a lesser extent, reactants derived from petroleum and other sources have been evaluated.

7.1.1 Conversion of α,β-Unsaturated Aldehydes

Among this type of reactant, most attention has been paid to CALD (Figure 5.2) and citral (Figure 5.3) for the reasons discussed earlier. Some information has been found on the conversion of crotonaldehyde.

Catalysis Series No. 37
Carbon Nanomaterials in Hydrogenation Catalysis
By Edward Furimsky

Published by the Royal Society of Chemistry, www.rsc.org

7.1.1.1 Cinnamaldehyde (CALD)

According to the mechanism of the conversion of CALD (Figure 5.2), the formation of three major products (CALC, HCALD and HCALC) is anticipated under controlled conditions. Optimized experimental conditions in the presence of a catalyst exhibiting suitable selectivity ensure high selectivity for the targeted products. Frequently, a high yield of CALC has been the objective. Otherwise, HCALD and HCALC are obtained in high yields. Of course, if the experimental conditions approach those encountered in HPR, the overall conversion ends up with the formation of hydrocarbons.

Ni *et al.*[2] studied Ru catalysts supported on CNTs in the HYD of CALD in cyclohexane. For the catalyst preparation, both nonfunctionalized CNTs and functionalized CNTs (CNT-f) were used. The activity of the catalysts was determined in an autoclave at 80 °C and 3 MPa of H_2. In this case, HCALD was the product of interest. Under these conditions, the Ru/CNT catalysts were more active, as indicated by 80% conversion of CALD with 72% selectivity with respect to HCALD. The higher activity of the Ru/CNT catalyst was attributed to more efficient dispersion of Ru on the support. In a study by Wang *et al.*,[3] commercial CNTs (od 10–29 nm, id 4–10 nm) were purified and oxidized by refluxing with HNO_3 at 140 °C for 14 h to obtain oxidized CNTs (OCNTs). To prepare an Ru(4%)/CNT catalyst with Ru particles confined in the interior of the CNTs, $RuCl_3$ was dissolved in acetone and added to the OCNTs. The suspension was treated ultrasonically followed by the evaporation of acetone. The resulting solid was dried at 110 °C and reduced in H_2 at 400 °C to obtain Ru/CNT catalysts with most of the Ru located in the interior of the CNTs. In the case of Ru/CNTa with Ru particles located on the exterior of the tubes, the OCNTs were treated ultrasonically with a solvent (*e.g.* acetone or toluene) before adding the acetone solution of $RuCl_3$. The mixture was heated at 60 °C with continuous stirring. Under these conditions, most of the acetone was evaporated. The resulting solid was treated as above. An Ru/CNTx catalyst was prepared by using a slurry of OCNTs in xylene mixed with an aqueous solution of $RuCl_{3-x}\cdot 3H_2O$. In addition, a solution of NH_4HCO_3 dissolved in NH_3xH_2O (26–28 wt%) was added in order to transfer OCNTs into the aqueous phase. Then, the same evaporation, drying and reduction procedures as for Ru/CNTa were adopted to obtain Ru/CNTx. The freshly reduced catalysts were treated at 700 °C for 2 h in a flow of N_2 to obtain Ru/CNTs-ht, Ru/CNTa-ht and Ru/CNTx-ht. The testing of these catalysts was conducted in an autoclave (70 °C, 1 MPa H_2) in the HYD of model reactants such as benzene, *p*-chloronitrobenzene and CALD. Figures 7.1 and 7.2[3] show that the heat treatment had an adverse effect on the conversion in the HYD of CALD but increased the selectivity with respect to CALC (Figure 5.2). Ma *et al.*[4] compared the activity of two Ru/CNT catalysts, namely one with Ru in the interior of the CNT channels (id 1–3 nm) and the other with Ru on the exterior of the CNTs. For the preparation of the former catalyst, the CNTs were evacuated and exposed to the vapor of an Ru precursor such as bis(2,4-dimethylpentadienyl)ruthenium at 120 °C for 48 h.

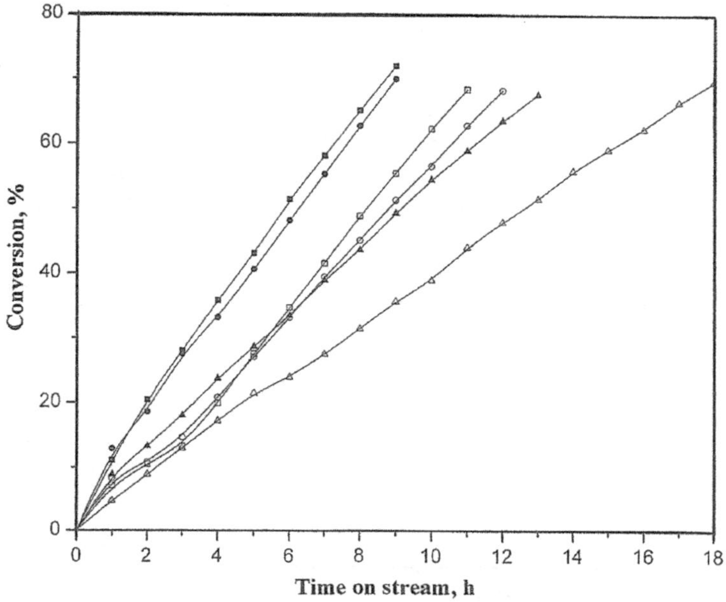

Figure 7.1 Effect of catalyst type on the conversion of CALD: ●, Ru/CNT; ■, Ru/CNTa; ▲, Ru/CNTx; ○, Ru/CNTs-ht; □, Ru/CNTa-ht; △, Ru/CNTx-ht. Reproduced from ref. 3 with permission from Elsevier, Copyright 2015.

The catalyst was then reduced in H_2 at 230 °C. In parallel, the Ru/CNT catalyst with Ru on the exterior was prepared by impregnation of CNTs with a solution of the same precursor dissolved in toluene. After removing the solvent, the catalyst was reduced as above. Both catalysts were used in the HYD of CALD in a batch system at 100 °C with 2 MPa of H_2 for 2 h. A significant difference between the selectivities of these catalysts was noted: the Ru/CNT catalyst with Ru located on the exterior exhibited high selectivity for the formation of CALC whereas similar amounts of CALC, HCALD and HCALC were present in the product obtained over the Ru/CNT catalyst with the Ru particles in the channels of CNTs. The difference may be attributed to the different geometry of the transition state at the active site comprising active metal with the adsorbed reactant.

Takenaka *et al.*[5] prepared two Pt/CNT catalysts, one with Pt located at the tips (edge regions) and the other with Pt on the external surface (basal planes) of CNTs. The former was prepared by the decomposition of ethylene in a mixture with H_2 on the surface of Pt/MgO catalyst at 973 K. Subsequently, the MgO was removed by treating the solid with HCl to obtain Pt(8.4%)/CNT catalyst. A Pt(15%)/CNT catalyst with Pt located on the external surface was prepared by impregnation of CNTs using butanol solution containing H_2PtCl_6. The performances of the catalysts were compared in the HYD of CALD at 313 K and 2 MPA of H_2 in an autoclave. Both catalysts exhibited high activity and selectivity with respect to CALC. In spite of a

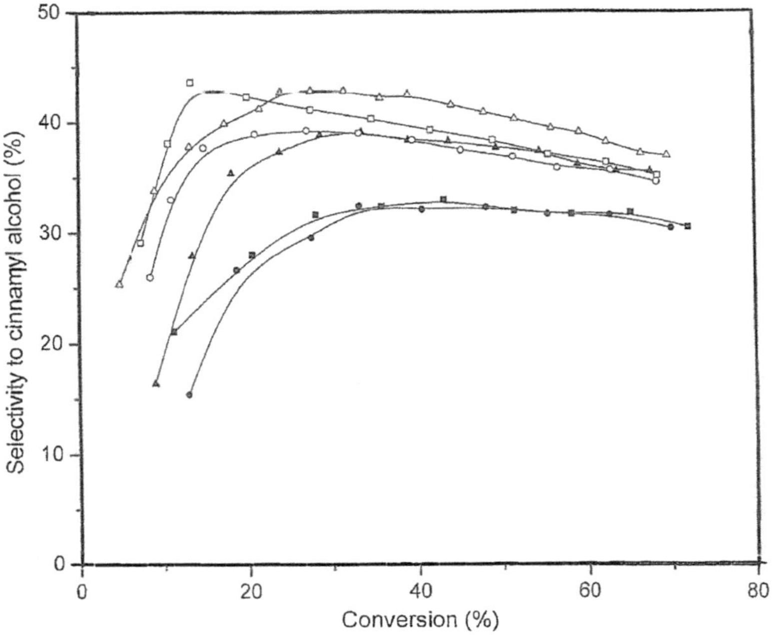

Figure 7.2 Effect of catalyst type on the selectivity with respect to CALC in the conversion of CALD: ●, Ru/CNTs; ■, Ru/CNTa; ▲, Ru/CNTx, ○, Ru/CNTs-ht; □, Ru/CNTa-ht; △, Ru/CNTx-ht.
Reproduced from ref. 3 with permission from Elsevier, Copyright 2015.

smaller content of Pt, the high activity and selectivity of the catalyst with Pt located at the tips of CNTs should be noted. This is perhaps the first study in which the effect of the location of active metals in CNTs on catalyst performance, *i.e.* tips *versus* external surface (basal plane), was investigated.

Liu *et al.*[6] used MWCNTs of varying external diameter as supports for the preparation of Pt(3%) catalysts to be used in the HYD of CALD. The supports were purified with nitric acid, washed and dried before loading with Pt *via* liquid-phase reduction. In this case, H_2PtCl_6 was used as the Pt precursor and formaldehyde as reductant. Most of the Pt was deposited on the exterior of the MWCNT supports with the Pt particle size approaching 11 nm. The catalysts were named in order of increasing outer diameter as Pt/MWCNT1, Pt/MWCNT2, Pt/MWCNT3 and Pt/MWCNT4. The textural properties of both the catalysts and supports are given in Table 7.1.[6] The testing was conducted in an autoclave using CALD dissolved in ethanol at 80 °C and an H_2 pressure of 2 MPa. All catalysts exhibited good activity, as indicated by an overall conversion approaching 80%. However, a significant difference in selectivity was noted. Thus, the selective HYD of the C=O bond was the main reaction over Pt/MWCNT1, Pt/ MWCNT2 and Pt/MWCNT3 catalysts, with the CALC selectivity approaching 60, 72 and 80%, respectively. Also, the HYD of the C=C bond yielding HCALD approached 65% over Pt/MWCNT4. The HYD of

Table 7.1 Surface areas and porosities of Pt(3%)/MWCNT catalysts.[6]

	Pt/MWCNT No.			
Property	1	2	3	4
Surface area/$m^2\,g^{-1}$				
BET	150.3	246.4	178.0	127.0
External	146.5	225.4	156.9	116.8
Total PV/$cm^3\,g^{-1}$	0.570	1.111	0.861	0.471
APD[a]/nm	14.2	17.1	18.5	13.5
	MWCNTs			
Diameter/nm				
Outer	8–15	10–20	20–30	>50
Average inner	6.5	6.5	7.5	6.8
APD[a]/nm	17.2	19.8	20.0	14.2
Surface area/$m^2\,g^{-1}$				
BET[a]	129.6	161.4	134.7	73.2
External	118.1	43.5	115.4	60.6
Total PV[a]/$cm^3\,g^{-1}$	0.583	0.892	0.803	0.284

[a]PV, pore volume; APD, average pore diameter.

both C=C and C=O bonds yielding HCALC was less than 30%. It was noted that no products other than CALC, HCALD and HCALC were detected. The unique selectivity of the Pt/MWCNT4 catalyst was attributed to the different surface acidities and curvatures of the outmost tubes. As already indicated, the activity and selectivity of the CNT-supported catalysts depend on the location of the active metals nanoparticles, *i.e.* in channels *versus* on the exterior of CNTs.

Tessonnier *et al.*[7] reported that an impregnation method using an aqueous solution of a Pd salt was suitable for the location of Pd nanoparticles (4–6 nm) in the channels of CNTs. Complete filling was facilitated by a low surface tension of the solvent. After drying and thermal treatment, a homogeneous Pd distribution in the channels was achieved. In the HYD of CALD, the Pd/CNT catalyst exhibited high activity for the HYD of the C=C bond, yielding HCALD. Under the same conditions, the activity of a high surface area commercial Pd/AC catalyst was much lower. The selective HYD of the C=C bond over a Pd/CNT catalyst yielding HCALD was also observed in a study by Bhanja *et al.*,[8] while CALC was the main product over a Pt/CNT catalyst. The effect of either Ni or Co on the activity and selectivity of Pt/CNT catalysts was investigated by Tian *et al.*[9] in the HYD of CALD. In this case, the CNT support was first purified and then doped with nitrogen in the reaction with NH_3 at 180 °C for 12 h. Under similar conditions, CNTs were treated with H_2O_2 and the treatment with NH_3 was avoided. The CNT support was impregnated using an aqueous solution of H_2PtCl_6 followed by the addition of $NaBH_4$ solution with vigorous stirring. The solid was filtered off, washed and dried at 70 °C. The activity was evaluated in a batch reactor by dispersing the catalyst in ethanol before the addition of CALD. The testing was conducted at 70 °C and 2 MPa of H_2. To tailor the selectivity, bimetallic Pt–Ni

Table 7.2 Activity and selectivity of catalysts in the hydrogenation[a] of CALD.[9]

Catalyst	Conversion/%	Selectivity/%			
		CALC[b]	HCALD[b]	HCALC[b]	ACE[b]
Pt–Co/CNT	85.4	88.4	6.6	5.0	0.3
Pt–Ni/CNT	82.9	25.2	60.6	13.6	0.6
Pt–Co/NCNT	99.7	87.9	4.1	7.7	0.3
Pt–Ni/NCNT	99.6	11.4	58.1	32.0	0.5
Pt–Co/OCNT	99.3	72.1	3.5	4.3	0.1
Pt–Ni/OCNT	97.9	27.4	38.7	32.7	1.2

[a]Conditions: 8.2 mmol CALD, 19.0 mL C_2H_5OH, 0.1 g catalyst, 70 °C.
[b]CALC, cinnamyl alcohol; HCALD, hydrocinnamaldehyde; HCALC, hydrocinnamyl alcohol; ACE, acetal.

and Pt–Co catalysts supported on either a nitrogen- or oxygen-modified support, *i.e.* NCNT and OCNT, respectively, were prepared. Table 7.2[9] shows that CNT modifications with both nitrogen and oxygen increased the conversion but had little effect on the selectivity with respect to CALC. There was a difference between the promoting effects of Ni and Co. A higher HYD activity of Ni resulted in a lower selectivity with respect to CALC. Concerning the activity and selectivity with respect to CALC, the Pt–Co/NCNT catalyst exhibited the best performance.

Monometallic Pt and Co catalysts and a bimetallic Pt–Co catalyst prepared by the atomic metal deposition method[10] on an MWCNT support were compared in the HYD of CALD by Wang *et al.*[11] For catalyst preparation, an organometallic complex such as (MeCp)$PtMe_3$ was used as the Pt precursor in a fluidized bed of MWCNTs using O_2 as the reactant.[12] The primary objective was the determination of the catalyst composition and test conditions to achieve high selectivity with respect to CALC (Figure 5.2). The tests were conducted at 80 °C and 1 MPa of H_2 in a batch reactor using CALD dissolved in 2-propanol. Under these conditions over a Pt–Co/CNT catalyst, the conversion of CALD reached 93% with a selectivity with respect to CALC of 93%. The activity and selectivity of the monometallic Pt/CNT and Co/CNT catalysts were much lower. DFT calculations conducted in parallel with the experiments revealed that the selectivity increase resulted from the activation of the C=O bond by Pt–Co nanoparticles enhanced by electronic effects of the MWCNT support.

The bimetallic Pt(0.36%)–Ni/CNT catalyst prepared by Li *et al.*[13] exhibited superior activity in the HYD of CALD. Prior to catalyst preparation, the CNT support was purified with nitric acid at 80 °C for 16 h. After filtering, washing and drying, the CNTs were impregnated with an aqueous solution of $Pt(NH_3)_4Cl_2$ to prepare a Pt/CNT catalyst. After impregnation, the solid was reduced using KBH_4. Co-impregnation with aqueous solutions of $Pt(NH_3)_4Cl_2$ and $Ni(NH_3)_4Cl_2$ was used to prepare a bimetallic Pt–Ni/CNT catalyst. The activities of these catalysts was evaluated in an autoclave at 70 °C and 2 MPa of H_2. In this case, the selectivity with respect to HCALD rather than to CALC was the objective. In parallel, similar catalysts but supported on graphite were prepared for comparison with the CNT-supported catalysts. In all respects, superior activity of the Pt(0.36%)–Ni/CNT catalyst was observed.

The effect of the inner diameter of the CNTs on the activity of Pt(3%)/CNT catalysts was investigated by Ma *et al.*[14] For this purpose, CNTs with id <10 nm and 60–100 nm were used. During impregnation, Pt particles were deposited on the outside of the former CNTs whereas they entered the interior of the CNTs with id 60–100 nm. The activity and selectivity of the catalysts were evaluated in the HYD of CALD as a model reaction. The tests were conducted in an autoclave at 80 °C and 3 MPa of H_2. Both catalysts exhibited high activity for the overall CALD conversion, but their selectivities were very different. Thus, complete HYD of both C=O and C=C bonds of the reactant over the catalyst with id <10 nm was observed, whereas the catalyst with id 60–100 nm was very selective for the HYD of the C=O bond.

Li *et al.*[15] used the HYD of CALD to CALC for comparison of the activities of monometallic Pt and bimetallic Co–Pt and Ni–Pt catalysts supported on CNTs. Higher activity of the bimetallic catalysts was indicated by conversions of CALD exceeding 90% and almost 70%, respectively, compared with 35% for the monometallic catalyst. Because of the mild conditions (343 K, 2 MPa H_2, 1.5 h, batch system), CALC was the main product rather than hydrocarbons. Other transition metals (*e.g.* Cr, Mn, Cu and Fe), in combination with Pt had an adverse effect on the performance of the catalysts. Qiu *et al.*[16] used monometallic Pd and Ru catalysts and bimetallic Pd–Ru catalysts supported on CNTs to study the HYD of CALD to CALC at 120 °C and 5 MPa of H_2 in an autoclave. The catalysts were prepared by the incipient wetness impregnation technique using an aqueous solution of either Pd nitrate or Ru chloride. Raw carbon nanotubes (RCNTs) were prepared by the arc discharge method and subsequently purified by refluxing in 65% HNO_3 at 120 °C for 8 h to obtain purified carbon nanotubes (PCNTs). Both CNTs were used as supports. The results are given in Table 7.3.[16] The performance of the

Table 7.3 Overall conversion and selectivity with respect to CALC in the HYD of CALD.[a] [16]

	Reaction time/min					
Catalyst	30	60	90	120	180	240
Pd/RCNT						
Conversion/%	10.3	11.3	13.1	13.9	16.7	20.5
Selectivity/%	21.3	21.2	19.9	20.1	19.2	18.0
Pd/PCNT						
Conversion/%	5.7	7.4	17.3	22.2	28.3	32.3
Selectivity/%	50.9	37.1	22.5	20.0	19.9	20.0
Ru/PCNT						
Conversion/%	6.7	10.8	16.7	20.8	—	—
Selectivity/%	78.7	78.7	78.6	75.0	—	—
Pd–Ru/PCNT						
Conversion/%	7.3	14.0	22.8	26.6	37.9	56.6
Selectivity/%	79.2	74.3	67.5	60.9	44.1	32.3

[a]RCNT, raw CNT; PCNT, purified CNT.

bimetallic catalyst was much better, as indicated by the higher conversion and selectivity with respect to CALC.

Li *et al.*[17] treated CNTs with nitric acid at 80 °C for 16 h to remove impurities and to introduce O-containing surface groups before using the CNTs as supports for Pt(0.5%)–Co(0.17%)/CNT bimetallic catalysts. Several methods of reduction were compared. In one method, the CNT support was co-impregnated with aqueous solutions of $Pt(NH_3)_4Cl_2$ and $Co(NH_3)_4Cl_2$, followed by reduction using HCHO at 70 °C for 6 h (Pt–Co/CNT1). Other reduction methods used H_2 at 400 °C for 4 h and KBH_4 at 80 °C for 2 h, yielding Pt–Co/CNT2 and Pt–Co/CNT3, respectively. Pt–Co/CNT5 catalyst was prepared by co-impregnation with aqueous solutions of H_2PtCl_6 and $Co(NO_3)_2$ and reduction with KBH_4. Pt–Co/CNT4 catalyst was prepared by sequential impregnation with an aqueous solution of $Co(NO_3)_2$ followed by an aqueous solution of $Pt(NH_3)_4Cl_2$. This solid was reduced with KBH_4. The catalysts were used in the HYD of CALD to CALC as targeted products (batch, 70 °C, 2 MPa H_2, 1.5 h). Table 7.4[17] shows that the method of reduction had a pronounced effect on catalyst performance. Thus, Pt–Co/CNT1 prepared by reduction with HCOH exhibited low activity. Reduction in H_2 improved the catalyst performance, but the yield of unwanted products such as acetal (ACE) and others was still too high. The best reducing agent was KBH_4, yielding very active catalysts, *i.e.* Pt–Co/CNT3–5 with high selectivity with respect to the product of interest such as CALC. It was evident that the monometallic Pt and Co catalysts exhibited low activity.

Monometallic Pt and Ru catalysts and also bimetallic Pt–Ru catalysts supported on SWCNTs and MWCNTs were compared with the CNF and AC analogs in the HYD of CALD to CALC (Figure 5.2). The bimetallic catalysts were prepared by two different methods: in one case the supports were co-impregnated (Pt–Ru) and in the other they were impregnated sequentially, first with Pt and then with Ru (Pt/Ru). After drying, the catalysts were reduced at 350 °C in H_2. The tests were conducted in an autoclave at 100 °C and 2 MPa of H_2 for 2 h using a solution of CALD in 2-propanol. The results

Table 7.4 Effect of preparation method on conversion and selectivity in the HYD of CALD over Pt–Co/CNT catalysts.[17]

		Selectivity/%				
Catalyst	Conversion[a]/%	HCALD	HCALC	CALC	ACE	Others
Pt–Co/CNT						
1	54.7 (3.5)[b]	—	—	6.3	—	93.7
2	96.7 (48.7)	2.0	4.8	13.6	30.0	49.6
3	87.4 (87.4)	4.7	2.9	92.4	—	—
4	85.9 (85.9)	4.7	2.1	93.3	—	—
5	85.7 (85.7)	21.5	3.3	75.3	—	—
Pt/CNT	48.2 (48.2)					
Co/CNT	11.5 (0)	—	—	—	—	100

[a]Conditions: 8.0 mmol CMA, 19.0 mL ethanol, 70 °C, 2 MPa, 1.5 h.
[b]Hydrogenation conversion: the sum of the conversion to HCALD, HCALC, CALC and ACE.

Table 7.5 Conversion of CALD and selectivity with respect to CALC over catalysts containing 2% of metals.[18]

	Conversion/selectivity[a]/%			
Support	Pt	Ru	Pt–Ru	Ru/Pt[b]
SWNT	85/28	65/31	58/42 (60/69)[c]	—
MWCNT	95/32 (97/66)	66/35	44/55 (79/93)	80/22
CNF	96/14 (90/15)	78/18	60/29 (92/17)	76/21
AC	20/62 (12/60)	22/51	15/64 (16/31)	23/38

[a]Conditions: 100 °C, 2 MPa H_2, 2 h.
[b]Sequential impregnation.
[c]Results in parentheses are for catalysts treated at 973 K in N_2.

in Table 7.5[18] show a much lower activity of the AC-supported catalysts. In terms of conversion, the activities of the monometallic and bimetallic catalysts were similar, but the selectivity of the latter with respect to CALC was much higher. Heat treatment at 973 K under N_2 improved the performance of the Pt–Ru/MWCNT catalyst. A high conversion but low selectivity with the CNF catalysts should be noted. The activity of the AC catalysts was much lower than that over the CNT-supported catalysts.

Another study on the HYD of CALD to HCALD over bimetallic Pt–Ni/CNT catalyst was reported by Li and co-workers.[19,20] The performance of this catalyst was significantly enhanced in the presence of a trace amount of a base or acid promoter. For example, 92% conversion of CALD with 96% selectivity with respect to HCALD was obtained in the presence of NaOAc.

The performances of monometallic Pt/CNT and bimetallic Pt–Fe/CNT catalysts in the conversion of CALD to CALC were compared by Dai *et al.*[21] Nitric acid-treated CNTs were used for the preparation of both catalysts using $Pt(acac)_2$ and $Fe(acac)_3$ as metal precursors. The reactions were conducted in a batch reactor at 60 °C and 2 MPa of H_2. For testing, CALD and catalyst in various ratios of were slurried with water in the reactor. In parallel, tests were conducted using organic solvents such as cyclohexane, ethyl acetate, *n*-hexane, *n*-heptane and 2-propanol instead of water. Under these conditions, water was the best reaction medium. For the bimetallic catalyst, the selectivity with respect to CALC exceeded 95% at a conversion of about 63% with a test duration of 0.5 h. The same selectivity was observed at almost complete CALD conversion at 1 h. A series of experiments performed in D_2O confirmed the involvement of D during the HYD of CALD. It was reported that a noncatalytic transfer of hydrogen from water to organic reactants required near-critical conditions of water, *e.g.* the critical temperature of water approaches 370 °C and a pressure of ~22 MPa. Nevertheless, almost certainly, the transfer of D under such mild conditions was facilitated by the catalyst surface. This rather unusual observation justifies further investigation of the participation of water in HYD.

Yan *et al.*[22] developed high-performance Pt(3.4 wt%)–Co(1.3 wt%)–Fe(2.6 wt%)/CNT trimetallic catalysts using an iterative response surface methodology. The performance of the catalysts was evaluated in the HYD of CALD with the objective of maximizing the yield of CALC. After nitric acid pretreatment, the CNTs were impregnated with an aqueous solution containing active metal precursors such as H_2PtCl_6, $Co(NO_3)_2$ and $Fe(NO_3)_3$. After sonication, the mixture was dried at 373 K overnight. The size of the active metal particles on the catalyst was less than 7 nm. Testing was carried out in an autoclave at 333 K and 1 MPa of H_2 for 1 h. With respect to the overall conversion, Pt was the dominant metal and Fe had a significant effect on selectivity. Under optimized conditions, the CALD conversion approached 86% with a yield of CALC exceeding 70%. The rarely used noble metal Ir was compared with Pt as part of MWCNT-supported catalysts by Machado *et al.*[23] Organometallic precursors such as $[Pt(CH_3)_2(C_8H_{12})]$ and $[Ir(\mu\text{-}SC(CH_3)_3)(CO)_2]_2$ were used for the preparation of the Pt and Ir catalysts, respectively. The precursor and nitric acid-treated MWCNTs were mixed in *n*-hexane under an inert atmosphere and kept for 24 h at 318 K with vigorous stirring. After filtering and drying, the catalysts were calcined under N_2 and reduced in H_2. For some catalysts, a partial reduction treatment in N_2 at 973 K was conducted to remove the remaining O-containing groups to obtain Ir/MWCNT973 and Pt/MWCNT973 catalysts. These catalysts were compared in the HYD of CALD in a batch reactor at 363 K and a total pressure of 1 MPa. Heptane was used as a solvent to avoid acetal formation, which is evident in alcohol solvents. The results of these tests are given in Table 7.6.[23] The thermal treatment at 973 K had a pronounced effect on catalyst selectivity. Thus, for the Pt catalysts, the untreated catalysts were selective for the formation of HCALD whereas the treated catalysts were selective for that of CALC. Both the treated and untreated Ir catalysts exhibited high selectivity for the formation of CALC.

Undoped and N-doped CNTs were used as supports for the preparation of Pd/CNT catalysts by Amadou *et al.*[24] For the preparation of the CNTs, ethane and NH_3 were used as the sources of carbon and nitrogen, respectively. Both Pd/CNT catalysts were used for the HYD of CALD dissolved in dioxane. The tests were conducted in a batch reactor at 353 K with bubbling H_2.

Table 7.6 Product distribution in the HYD of CALD over Ir and Pt catalysts supported on MWCNTs.[23]

		Selectivity/%			
Catalyst[a]	TOF[b]/s^{-1}	CALC	HCALD	HCALC	Others
Ir(1%)/MWCNT	1.5	57	18	13	12
Ir(1%)/MWCNT973	1.1	54	19	14	13
Ir (2%)/MWCNT	0.8	56	18	10	16
Ir(2%)/MWCNT973	1.4	68	17	11	4
Pt (1%)/MWCNT	0.9	8	51	13	28
Pt(1%)/MWCNT973	1.1	68	10	9	13
Pt(3%)/MWCNT	1.5	20	33	21	26
Pt(3%)/MWCNT973[c]	6.6	45	18	24	13

[a]Conditions: 363 K, total pressure 1 MPa.
[b]TOF, turnover frequency.
[c]74% conversion of CALD.

For Pd/CNT supported on N-doped CNTs, the CALD conversion was almost twice that over the undoped Pd/CNT catalyst with selectivity with respect to HCALD approaching 90% compared with less than 80% for the latter catalyst. Also, over Pd/AC, the selectivity decreased from 60% to less than 30% with increasing CALD conversion.

7.1.1.2 Conversion of Citral

With respect to the reactant reactivity, the presence of α,β-conjugated C=C and C=O double bonds in citral indicates some similarity in the mechanism of the HYD of CALD, although the structural difference due to the presence of an aromatic ring in the latter may modify the reaction pathways. Nevertheless, suitable selectivity of the catalysts and optimal conditions are required to maximize the yield of the targeted products as shown in the mechanism in Figure 5.3.

Frequently, nerol and geraniol are the products of interest. Two types of CNTs, *i.e.* one prepared by arc discharge and the other by CVD, were compared with CB as the support for Pt catalysts to be used for the HYD of citral at 323 K and 5 MPa of H_2 in an autoclave.[25] Both catalysts contained 2 wt% of Pt. With respect to the conversion and selectivity with respect to nerol and geraniol, superior performance was observed over the Pt catalyst supported on CNTs prepared by the CVD method. The HYD activity and selectivity of the Pt/CB catalyst were much lower than those of the Pt/CNT catalyst.

Wang *et al.*[26] prepared Pt catalysts supported on CNTs, SiO_2 and Al_2O_3 using an atomic layer deposition method.[10,27,28] The activity and selectivity of the catalysts were evaluated in the HYD of citral in a batch reactor at 70 °C and 1 MPa of H_2. Citral was dissolved in 2-propanol. The selectivity was determined from the yield of unsaturated alcohols. Pt/SiO_2 exhibited the best performance, with 82% conversion of citral at 58% selectivity. For the Pt/CNT catalyst, the selectivity approached almost 70% at 40% conversion. However, Pt/CNT catalyst was much more stable than the Pt/SiO_2 catalyst; little change in activity and selectivity was observed during six repeated cycles. At the same time, a decrease in the activity and selectivity of the Pt/SiO_2 catalyst was evident already after the first repeated cycle.

Guo *et al.*[29] reported that the morphology of Pt particles supported on poly(acrylic acid)-grafted MWCNTs played a significant role in the HYD of citral (333 K, 1.0 MPa H_2, 2 h, batch process). It was observed that during the impregnation of the support, the molar ratio of KNO_3 to H_2PtCl_6 was an important parameter. Without KNO_3, irregular Pt particles of size 5–7 nm were obtained. With the addition of KNO_3, larger single-crystal Pt nanoparticles were formed. The crystal shape changed from tetra- and octahedral to tetra- and octapods on increasing the molar ratio of KNO_3 to H_2PtCl_6 from 5.5 to 11. The activity of catalysts with irregularly shaped nanoparticles exceeded 90% with selectivity with respect to mono-HYD approaching 96%. For the polyhedrons and polypods of Pt nanoparticles supported on

MWCNTs, the conversion was 45.0 and 40.4%, respectively, In comparison, less than 25% citral conversion was achieved over a Pt/AC catalyst.

Ananthan *et al.*[30] used monometallic Pt and Ru and bimetallic Pt–Ru catalysts supported on CNTs for the HYD of citral at 160 °C in an autoclave. In all respects, the bimetallic catalysts exhibited better performance. Before the addition of active metals, the CNTs were first purified/functionalized with nitric acid treatment, followed by impregnation using aqueous solutions of Pt and Ru precursors such as $H_2PtCl_6 \cdot 6H_2O$ and $RuCl_3 \cdot 6H_2O$, respectively, mixed with ethylene glycol. Prior to testing, the catalysts were thermally treated at either 375 or 675 °C. Preoxidation followed by thermal treatment increased the activity and selectivity of the catalysts. Over a bimetallic Pt–Ru/CNT catalyst, the citral conversion approached 85% and the selectivity with respect to unsaturated alcohols reached 95%.

The effect of purification of commercial MWCNTs to be used as supports for Pt/CNT catalysts was studied by Zgolicz *et al.*,[31] involving successive treatments with 10 wt% aqueous solutions of HCl, HNO_3 and HF at room temperature for 48 h. Subsequently, the CNTs were filtered, washed with deionized water and dried at 120 °C for 24 h to obtain a CNT-CNF support. In another treatment, the CNTs were treated with hot HNO_3 at 120 °C for 3 h to obtain a CNT-N support. In addition, CNTs were treated ultrasonically with citric acid. The paste obtained after solvent evaporation was dried at 120 °C and heat treated at 300 °C in a flow of N_2 (CNT-C support). The supports were treated at 700 °C in either H_2 or N_2. The catalysts were prepared by impregnation of the supports to obtain 5 wt% Pt. Subsequently, the catalysts were reduced at 350 °C in a flow of H_2 for 3 h. Some catalysts were treated at 700 °C in either H_2 or N_2. The catalysts were tested in the HYD of citral to give unsaturated alcohols (UAs) in a discontinuous system with an arrangement for periodic sample removal. The results in Table 7.7[31] show the effect of the preparation methods on the catalyst performance. The superiority of the Pt/CNT and Pt/CNT-C catalysts compared with the Pt/CNT-CNF and

Table 7.7 Reaction time to reach 95% citral conversion and selectivity with respect to unsaturated alcohols (UAs).[31]

Catalyst[a]	Reaction time/h	Selectivity for UAs/%
Pt/CNT-CNF	7.0	7.2
Pt/CNT-CNF-r	3.0	10.6
Pt/CNT-CNF-t	1.0	26.8
Pt/CNT-N	11.0	10.2
Pt/CNT-N-r	7.5	20.1
Pt/CNT-N-t	1.5	28.0
Pt/CNT-C	0.8	83.0
Pt/CNT-C-r	1.4	81.3
Pt/CNT-C-t	1.7	91.1
Pt/CNT	1.2	84.2
Pt/CNT-r	1.7	81.5
Pt/CNT-t	4.2	94.2

[a]-r, reduced in H_2 at 700 °C; -t, thermally treated in N_2 at 700 °C.

Pt/CNT-N catalysts is clearly evident. It was observed that the treatment with mineral acids removed impurities from the CNTs, including Fe. However, Fe in combination with Pt beneficially influenced the activity of the Pt/CNT and Pt/CNT-C catalysts. The CNT pretreatment with citric acid had little effect on Fe removal. Thermal treatment of these catalysts at 700 °C under N_2 further improved the selectivity with respect to unsaturated alcohols.

MWCNTs prepared by the CVD method were used as supports by Qin *et al.*[32] for the preparation of Pt catalysts. In addition, the same supports were functionalized by deposition of poly(acrylic acid) using a grafting method to obtain MWCNT-g, which was used to prepare a Pt/MWCNT-g catalyst. For Pt deposition, the supports were mixed with a solution of Pt precursor in ethylene glycol and sonicated, followed by magnetic stirring at 160 °C for 30 min, filtration and drying. In parallel, a Pt/AC catalyst was prepared by conventional wetness impregnation. The activities of the catalysts were compared in an autoclave (80 °C, 2 MPa, 2 h) using citral as a model reactant dissolved in hexane. The following activity order under the same conditions was established: Pt/MWCNT-g > Pt/MWCNT > Pt/AC, or in terms of conversion 87.2, 54.4 and 29.9%, respectively.

In a study by Stassi *et al.*,[33] monometallic Pt ands bimetallic Pt–Fe, Pt–Sn, Pt–In and Pt–Ga catalysts supported on nitric acid-treated CNTs were prepared by two methods, *i.e.* conventional impregnation and the deposition–reduction method. In the former method, CNTs were impregnated using an aqueous solution of H_2PtCl_6, whereas bimetallic catalysts were prepared by successive impregnation of the Pt monometallic solid with $FeCl_3$, $SnCl_2$, $In(NO_3)_3$ and $Ga(NO_3)_3$ precursors. After impregnation, the catalysts were reduced at 350 °C in H_2 for 3 h. Subsequently, the catalysts were thermally treated in N_2 at 700 °C for 12 h. Preparation *via* the deposition–reduction method involved impregnation with an aqueous solution of H_2PtCl_6, followed by reduction with $NaBH_4$ in NaOH. After drying, the solid was thermally treated as above. The catalysts were compared in the HYD of citral. Regarding the activity and selectivity with respect to unsaturated alcohols, the best performance was exhibited by Pt–Sn/CNT and Pt–In/CNT catalysts prepared by impregnation followed by treatment in N_2. The catalysts prepared by the deposition–reduction method were less active.

7.1.1.3 *Conversion of Crotonaldehyde*

Lordi *et al.*[34] studied the conversion of prenal (3-methylcrotonaldehyde; 3-methyl-2-butenal) to prenol (3-methyl-2-butenol; an unsaturated alcohol) over a Pt(10%)/CNT catalyst. The SWCNT support was prepared by the arc discharge method. The solid product was purified and oxidized by nitric acid treatment before being used for catalyst preparation. Pt deposition involved mixing the oxidized CNTs with K_2PtCl_4 dissolved in ethylene glycol followed by centrifuging, washing and drying. The tests were conducted in a continuous system at temperatures ranging from room temperature to 110 °C in a flow of a near atmospheric pressure H_2. Desirable selectivity with respect to prenol was achieved at 75 °C.

7.1.2 Summary of Studies on Conversion of α,β-Aldehydes

Table 7.8 summarizes studies on the conversion of the α,β-unsaturated aldehydes such as CALD and citral over noble metal catalysts supported on CNTs. The mildness of the experimental conditions employed in these studies should be noted. This suggests that complete conversion and/or HDO to hydrocarbons was not an objective of these studies. During the conversion of CALD and citral, unsaturated alcohols such as CALC (Figure 5.2) and also geraniol and nerol (Figure 5.3), respectively, were the targeted products in most of these studies. This suggests that the selection of catalysts and experimental conditions is a delicate process for maximizing the yields of these products. To various extents, under the conditions applied, unwanted HDO and associated water formation could be observed. Potentially, water can interact with active metals and thus obstruct catalytic reactions. Therefore, catalyst deactivation during the overall conversion of

Table 7.8 Noble metal-containing catalysts supported on CNTs for conversion of CALD and citral in batch reactors.

Catalyst	Conditions	Ref.
Conversion of CALD		
1. Ru/CNT	353 K; 3 MPa; cyclohexane	2
2. Ru/CNT	343 K; 1 MPa; 2-propanol	3
3. Ru/CNT	353 K; 2 MPa; 2-propanol	4
4. Pt/CNT	313 K; 2 MPa; ethanol + water	5
5. Pt/CNT	353 K; 2 MPa; ethanol	6
6. Pd/CNT; Pd/AC	353 K; 1 atm; dioxane	7
7. Pd/CNT; Pt/CNT		8
8. Pt–Ni/CNT; Pt–Co/CNT	343 K; 2 MPa; ethanol	9
9. Pt/CNT; Co/CNT; Pt–Co/CNT	353 K; 1 MPa; 2-propanol	10–12
10. Pt/CNT; Pt–Ni/CNT	343 and 353 K; 2 MPa; ethanol	13, 15
11. Pt/CNT; Pt–Ni/CNT	343 and 353 K; 2 MPa; ethanol	17, 19
12. Pt/CNT; Pt–Co/CNT	353 K; 3 MPa	14
13 Pd/CNT; Ru/CNT; Pd–Ru/CNT	393 K; 5 MPa; ethanol	16
14. Pt/CNT; Ru/CNT; Pt–Ru/CNT; Pt–Ru/AC	373 K; 2 MPa; propanol	18
15. Pt/CNT; Pt–Fe/CNT	333 K; 2 MPa; water, *n*-hexane, cyclohexane, ethyl acetate, *n*-heptane, 2-propanol	20
16. Pt–Co–Fe/CNT	333 K; 1 MPa; ethanol	22
17. Pt/CNT; Ir/CNT	363 K; 1 MPa; heptane	23
18. Pd/CNT; Pd/AC	353 K; 1 atm H_2; dioxane	24
Conversion of citral		
19. Pt/CNT; Pt/CB	323 K; 5 MPa; 2-propanol	25
20. Pt/CNT; Pt/SiO_2; Pt/Al_2O_3	343 K; 1 MPa; 2-propanol	26
21. Pt/CNT; Pt/AC	333 K; 1.0 MPa	29
22. Pt/CNT; Ru/CNT; Pt–Ru/CNT	433 K	30
23. Pt/CNT	343 K; 1 atm; 2-propanol	31
24. P t/CNT; Pt/AC	353 K; 1 MPa; hexane	32
25. Pt–Fe/CNT; Pt–Sn/CNT; Pt/CNT; Pt–In/CNT; Pt–Ga/CNT	343 K; 1 atm; 2-propanol	33

CALD and citral to hydrocarbons caused by water may not be avoided completely. Deactivation by coke deposition is of lesser importance because of the mild conditions. To minimize catalyst deactivation, a delicate balance involving temperature, H_2 pressure and structure of the catalyst surface must be established. This was confirmed in several studies in which the original catalyst activity could be maintained during several subsequent cycles while maintaining high selectivity. The same was not observed for catalysts supported on amorphous carbon supports such as ACs and CBs. When tested under identical conditions, noble metal catalysts supported on CNTs exhibited superior performance to the corresponding catalysts supported on oxidic supports (*e.g.* γ-Al_2O_3, SiO_2, TiO_2, SiO_2–Al_2O_3).

In spite of the low temperatures employed, a relatively high pressure of H_2 was used in the conversion of CALD and citral. It is believed that a sufficient H_2 pressure is required to minimize catalyst deactivation. Thus, a study by Ren *et al.*[35] indicated potential bonding of oxygen with noble metals, if eliminated from the reactants. Consequently, the involvement of active metals in catalysis is diminished.

Table 7.8 shows that among the noble metals, Pt either alone or in combination with other noble metals and/or transition metals has been used most frequently. To a lesser extent, Ru and Pd have been used. Of particular importance is the structure of the CNTs used for catalyst preparation. For both CALD and citral, the experimental conditions employed for testing were similar. Most of the studies were conducted in batch reactors rather than in continuous systems. Almost universally, functionalization of CNTs using hot nitric acid, prior to catalyst preparation, was used. However, the use of other oxidizing agents (*e.g.* H_2O_2, H_2SO_4, air) has also been noted. A wide range of CNT surface structures can be established when functionalization is combined with doping. This was evidenced by the difference in the activity and selectivity exhibited by the same active metals when supported on CNTs pretreated using a different method.

With respect to catalyst activity and selectivity, the medium in which the α,β-unsaturated aldehydes such as CALD and citral are dissolved for catalytic tests may play an important role.[20] In most of the studies listed in Table 7.8, only one solvent was used. However, Dai *et al.*[21] indicated the importance of the solvent selection for the overall efficiency of the process. Moreover, the ease of the isolation of products from the reaction mixture at the end of a test depends on the type of solvent. It is believed that the solvent selection requires additional attention for optimizing HYD catalysis.

7.1.3 Conversion of Reactants of Lignocellulosic Origin

Lignocellulosic biomass consists of three main components, cellulose, hemicellulose and lignin. Further depolymerization of these components leads to the formation of numerous reactants, as shown in Figure 5.1 for cellulose and hemicellulose. Some of these reactants have attracted attention as sources of organic chemicals and fuels. The abundance of lignin makes this component of lignocellulosic biomass an attractive source of numerous monomeric reactants that can be further upgraded to useful products. Frequently, noble metal-containing catalysts supported on carbon nanomaterials have been evaluated for these applications and less attention has been paid to other metals.

7.1.3.1 Glucose and Sorbitol

These two reactants are related as sorbitol can be produced *via* the HYD of the aldehyde group of D-glucose. This reaction may be considered as one of the routes for the gradual conversion of cellulose *via* glucose to sorbitol, glycols and finally hydrocarbons. Cellobiose is the dimer of glucose and may be the first intermediate during cellulose depolymerization. According to Figure 5.1, other intermediates formed during the hydrolysis of cellulose may be involved in the overall conversion to hydrocarbons. Of course, some of these intermediates are valuable products. Catalysts that exhibit high selectivity for such products have been a focus of attention. Sorbitol can be prepared directly from cellulose, as demonstrated by Ribeiro and co-workers.[36,37] For this purpose, they compared Ru(0.4%) catalysts supported on CNTs, ACs, graphite, CBs, Al_2O_3 and HY zeolite. The catalysts were prepared by incipient wetness impregnation of the supports using an aqueous solution of $RuCl_3$. The testing was conducted in an autoclave at 205 °C and 5 MPa of H_2 under aqueous conditions. The Ru/CNT catalyst exhibited the best performance, as indicated by the selectivity with respect to sorbitol approaching 60% at conversions higher than 75%. This catalyst was used for the conversion of waste cellulosic materials (*e.g.* wool, textiles, papers) directly to sorbitol. For the materials as received, conversions of up to 38% with selectivity with respect to sorbitol of 10% were achieved. Ball-milling of the wastes with Ru/CNT catalyst resulted in complete conversion, with selectivity with respect to sorbitol approaching 50%. In another study, Ribeiro *et al.*[38] used an Ru(0.4%)/CNT catalyst for the conversion of cellulose and xylan to sorbitol and xylitol, respectively. The highly crystalline cellulose was ball-milled to enhance the reactivity whereas xylan was used as received. The experiments were carried out in a batch reactor at 170 and 205 °C and 5 MPa of H_2. The feeds were tested both separately and in a mixture. For the mixture, the yields of sorbitol and xylitol were 76.5 and 46.3%, respectively. Compared with the experiments with single feeds, these yields are higher by 25 and 15%, respectively. This suggests that in the mixture, synergy between the feeds had a beneficial effect on the conversion and selectivity. Deng *et al.*[39a] investigated the conversion of cellulose to sorbitol over a series of Ru/CNT catalysts. For the catalyst preparation, nitric acid-treated CNTs were impregnated using an aqueous solution of $RuCl_3$. For the comparison of the activities and selectivities, catalysts containing Fe, Co, Ni, Pt, Rh, Ir and Ag, all supported on the same CNTs, were also used. The effect of the nitric acid concentration used for the pretreatment of CNTs on the acidity (Figure 3.1) and H_2 chemisorption of the Pd/CNT catalysts was investigated.

It was evident that the performance of the catalysts improved with increase in the concentration of nitric acid used for pretreatment. Before testing, cellulose samples with different degrees of crystallinity and polymerization were prepared by treatment with 85 and 43 wt% H_3PO_4. The experiments were conducted in an autoclave under aqueous conditions at 185 °C and 5 MPa of H_2 for 24 h. Figure 7.3[39b] shows that the selectivity of the Ru/CNT catalyst with respect to sorbitol was superior to that with all other catalysts. The yield of sorbitol increased with decreasing crystallinity of cellulose. Thus, for the untreated cellulose, the yield of sorbitol approached 36% compared with 69% for the pretreated cellulose.

To study the conversion of glucose to sorbitol, Aho *et al.*[40] used a commercial Ru(0.77%) catalyst on carbon extrudates (Ru/AC) and also Ru supported on pristine CNTs and N-doped CNTs, *i.e.* Ru/CNT and Ru/NCNT, respectively. The catalysts were prepared by treating the supports with $RuCl_3 + H_2O_2$ solution. In parallel, catalysts were prepared by a colloidal method (Ru/NCNT-Col). The Ru/NCNT-Col catalyst was thermally treated at 430 °C in either H_2 (Ru/NCNT-ColH_2) or Ar (Ru/NCNT-ColAr). The conversion of glucose to sorbitol was studied in a continuous system in the temperature range 90–130 °C and at a total pressure of 2 MPa. The conversions and selectivities in Figure 7.4[40] indicate the superiority of the Ru/NCNT catalysts. Table 7.9[40] shows that with respect to activity and selectivity, the size of the Ru particles may play a more important role than the total amount of Ru on the surface. It was evident that the performance of the Ru/AC catalyst was inferior to that of the CNT-supported catalysts.

The CNT supports in Table 3.1 and corresponding untreated CNT0 (od 9.5 nm, length 1.5 μm) were used for the preparation of Ru catalysts by incipient wetness impregnation using an aqueous solution of $RuCl_2$.[41] The HYD of glucose was performed in an autoclave at 205 °C and 5 MPa of H_2 for 1 and/or 5 h. During blank runs (with supports only), CNT2 was most active,

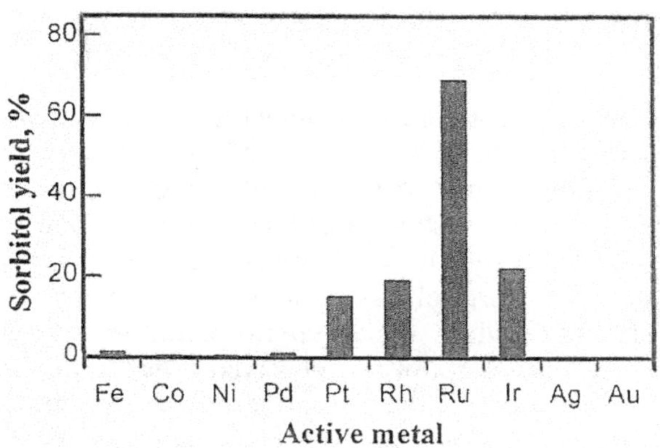

Figure 7.3 Effect of active metal type on conversion of cellulose to sorbitol. Reproduced from ref. 39*b* with permission from Elsevier, Copyright 2009.

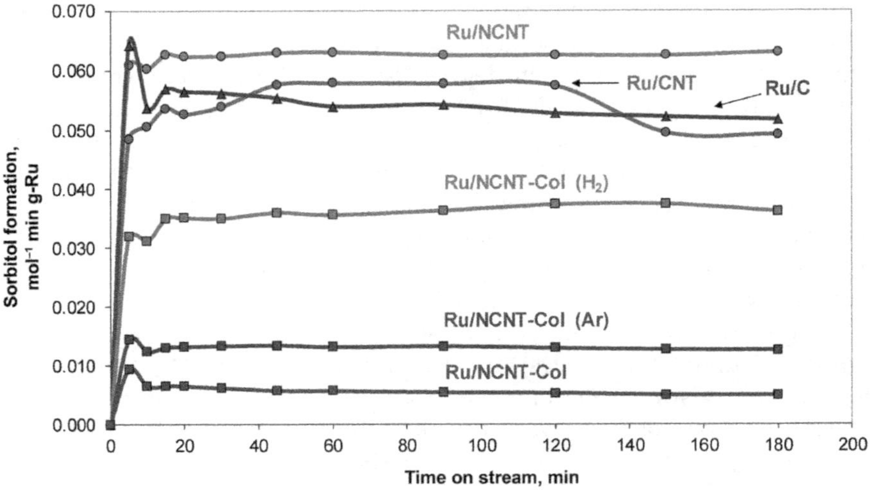

Figure 7.4 Effect of pretreatment of CNT supports on the conversion of cellulose to sorbitol over Ru/CNT catalysts.
Reproduced from ref. 40 with permission from the Royal Society of Chemistry.

Table 7.9 Conversion of glucose and selectivity with respect to sorbitol.[40]

Catalyst[a]	Ru content/%	Ru size/nm	Conversion/%	Selectivity[b]/%
Ru/NCNT	2.2	1.4	91.4	98.2
Ru/AC	0.77	2.9	41.5	98.7
Ru/CNT	3.1	1.4	65.0	97.4
Ru/NCNT-ColH_2	n/a	n/a	31.2	97.9
Ru/NCNT-ColAr	n/a	n/a	24.1	97.9
Ru/NCNT-Col	6.3	2.9	8.3	93.6

[a]Conditions: 130 °C, 20 bar, 0.2 mol L^{-1} aqueous glucose solution, flow rate 2.0 mL min^{-1}.
[b]Mannitol and fructose account for the difference from 100%.

giving a conversion of glucose of >60% but with <8% selectivity with respect to sorbitol. Almost complete conversion of glucose was achieved over Ru/CNT catalysts. After testing for 5 h, the following order of selectivity with respect to sorbitol was established: Ru/CNT2 > Ru/CNT2-900 > Ru/CNT), *i.e.* 56.8, 51.3 and 45.5%, respectively. However, the selectivity after 1 h was greater, suggesting that sorbitol was undergoing secondary reactions. One-pot hydrolytic HYD of cellulose was also performed. In this case, the highest selectivity with respect to sorbitol was observed over the Ru/CNT0 catalyst.

ACs and CNTs were compared as supports for monometallic Ru and Ni and also bimetallic Ru–Ni catalysts in the conversion of cellulose to sorbitol.[42] The catalysts were prepared by impregnation of the supports. In a typical experiment, a mixture of cellulose and catalyst in water was transferred to an autoclave, with reaction at 205 °C and 5 MPa of H_2 for 5 h.

Two types of experiments were performed, *i.e.* one in which cellulose and the catalyst were ball-milled together and the other in which cellulose and the catalyst were ball-milled separately and mixed together before the test. On the basis of the selectivity and yield of sorbitol, the bimetallic Ru–Ni/CNT catalysts exhibited the best performance. However, as the results in Table 7.10[42] show, high activity and selectivity were also observed over the Ru–Ni/AC catalyst prepared by ball-milling the support and metal precursor together.

Pan *et al.*[43] compared the activity of an Ru/MWCNT catalyst with that of Ru/Al_2O_3, Ru/SiO_2 and RANEY® Ni catalysts in the conversion of glucose to sorbitol in an autoclave (393 K, initial H_2 pressure 4 MPa, 120 min). The superior activity of the Ru/MWCNT catalyst was confirmed by the results in Table 7.11.[43] The catalyst preparation involved the impregnation of the support with an alcoholic solution of $RuCl_3$, followed by vacuum drying at 328 K for 5 h and heating at 573 K for 2 h in a flow of He. Before the experiments, the catalysts were reduced in a flow of H_2 at 673 K for 2 h.

Guo *et al.*[44] studied the effect of different functional groups (amino, carboxylic and hydroxyl) introduced to MWCNTs on the performance of corresponding Ru catalysts. In addition, a nitrogen-doped support was used for comparison. Ru(4%) catalysts were prepared by incipient wetness impregnation of the supports. After evaporation of water and drying, the catalysts were reduced in H_2 at 250 °C for 2 h. The catalysts were used for the hydrogenolysis of sorbitol to glycols. The experiments were conducted in an autoclave (210 °C, 6 MPa, 2 h) using a mixture of sorbitol in water, catalyst and an alkali metal hydroxide. Among several hydroxides, NaOH had the most beneficial effect on conversion and selectivity. The functional groups on the catalyst surface enhanced the anchoring of Ru particles and as such prevented their agglomeration. Sun *et al.*[45] compared the activities of bimetallic Ru–Mo/CNT and Ru–Mo/AC catalysts in the conversion of sorbitol to hydrocarbons. The former catalyst was more active and selective with respect to C_5 and C_6 alkanes. Thus, as Table 7.12[45] shows, at 280 °C over the

Table 7.10 Conversion of cellulose, selectivity and yield of sorbitol.[42]

	Run duration 1 h			Run duration 5 h		
Catalyst	Conversion/%	Selectivity/%	Yield/%	Conversion/%	Selectivity/%	Yield/%
Ru/AC	61.1	42.8	26.2	87.8	48.0	42.1
Ni/AC	41.8	15.0	6.3	77.9	8.9	6.9
Ru–Ni/AC	54.1	58.6	31.7	88.3	58.7	51.8
Ru/CNT	52.9	46.9	24.8	83.5	60.9	50.9
Ni/CNT	49.3	13.4	6.6	89.3	8.5	7.6
Ru–Ni/CNT	74.5	55.6	41.4	99.7	61.2	61.0
Ru–Ni/AC[a]	86.1	86.3	74.3	86.6	80.1	69.4
Ru–Ni/AC[b]	54.2	24.4	13.2	80.2	25.4	20.4
Ru–Ni/CNT[a]	99.3	71.3	70.8	100	64.4	64.4
Ru–Ni/CNT[b]	78.3	47.4	37.1	99.4	63.6	63.2

[a]Cellulose and catalyst were ball-milled together.
[b]In addition to cellulose, the catalyst was also ball-milled.

Table 7.11 Hydrogenation activity of Ru(5%) catalysts and RANEY® Ni.[43]

	Catalyst			
	Ru/MWCNT	Ru/Al_2O_3	Ru/SiO_2	RANEY® Ni
Conversion/%	62.5	47.9	41.7	27.1
Turnover frequency/h^{-1}	168.3	129.0	106.7	2.1

Table 7.12 Activity and selectivity of catalysts in the HYD of sorbitol.[45]

			Yield of hydrocarbons/%			
Catalyst[a]	Temperature/°C	Conversion/%	C_1	C_2–C_4	C_5	C_6
Ru–Mo/CNT	240	100	10.8	14.9	19.1	36.7
	280	100	13.7	20.5	21.5	39.8
Ru–Mo/AC	280	99	18.3	37.7	20.3	14.2
	300	99	21.0	30.8	18.6	27.8

[a]Conditions: WHSV = 1.59 h^{-1}, 4 MPa H_2.

Ru–Mo/CNT catalyst, the yield of the most desirable C_5 and C_6 alkanes approached 61% compared with about 35% over the Ru–Mo/AC catalyst. The tests were performed in a continuous fixed-bed reactor at 220–300 °C and 4 MPa using a solution of 20% sorbitol in water. These results confirmed that hydrocarbons are the final products if the testing conditions approach those typically employed in HPR.

7.1.3.2 Conversion of Xylose

Perez *et al.*[46] compared Ru, Pd, Pt, Rh and Au catalysts supported on MWCNTs in the conversion of xylose in a semi-batch system at 130 °C and 3 MPa of H_2 in water–2-propanol medium. For catalyst preparation, nitric-acid treated CNTs were impregnated using salts (metal chlorides) of the corresponding precursors to obtain 1 wt% of metal. With respect to the conversion to alcohols, the following activity order was established: Ru $\gg$ Pd $\gg$ Pt > Rh $\approx$ Au. As the main products, xylitol and furfuryl alcohol (FFA) accounted for almost all of the converted reactant.

Liang and Jiang[28] compared Pt/MWCNT catalyst with commercially available Pt/AC, Ru/AC and RANEY® Ni catalysts in the conversion of xylose to xylitol. The Pt/MWCNT catalyst was prepared by the atomic layer deposition method.[10,27] The experiments were conducted in an autoclave at 120 °C and 5.5 MPa of H_2 using an aqueous solution of xylose. In the first run, the performances of Pt/CNT and Pt/AC catalysts were similar. However, a steady decline in the activity of the commercial Pt/AC catalyst was observed in subsequent runs with spent–rejuvenated catalysts. At the same time, the Pt/CNT catalyst exhibited no decrease in activity even after 20 repeated runs. The activity loss with the Pt/AC catalyst was attributed to the leaching of Pt during the experiments. For example, for the Pt/CNT catalyst, the Pt content decreased from 5.4 to 5.2% (3.7% loss) after 20 repeated runs, whereas for the Pt/AC catalyst, the Pt content decreased from 3 to 2.2% (30% loss) after 10 cycles. The excellent performance of the Pt/CNT catalyst was attributed to the efficient dispersion of Pt metal on the CNT surface facilitated by the atomic layer deposition method. The Ru/AC and RANEY® Ni catalysts were much less active.

7.1.3.3 Furfural (FF) and Related Reactants

According to the mechanism in Figure 5.1, furfural structures can be obtained by the gradual hydrolysis of cellulose *via* glucose/fructose and xylose. Value-added products can be obtained by the HYD of the hydrolysis products/reactants. Pd and Ru catalysts supported on CNTs and CBs were used for the aqueous-phase HYD of FF.[47] It was observed that the adsorption of Pd on CNTs was less strong than on CBs. Consequently, Pd complexes supported on CNTs were reduced at a lower temperature, yielding highly dispersed nanoparticles of Pd. The Pd(1.5%)/CB catalyst exhibited a high selectivity for FFA at 50 °C and 0.5 MPa H_2. Under these conditions, the 1.5% Pd/CNT catalyst was not active but was more active at 90 °C with 0.5 or 2.0 MPa of H_2. During the aqueous-phase HYD of FF, the Ru catalysts exhibited lower activity than Pd catalysts.

Wang *et al.*[48] studied the HYD of FF to FFA over Pt/MWCNT catalysts in a batch reactor (423 K, 2 MPa, 5 h). Prior to Pt deposition, the MWCNT support was subjected to nitric acid pretreatment. The duration of the pretreatment influenced the textural properties of the support (Table 7.13) and also the activity and selectivity of the corresponding Pt/MWCNT catalysts (Table 7.14).[48] In these tables, MWCNT-1, -2 and -4 indicate the duration of the acid pretreatment in hours, *i.e.* the catalysts were treated for 1, 2 and 4 h. The oxygen groups created by the pretreatment ensured effective dispersion of Pt particles. This is supported by the increased conversion from 88 to 94% on increasing the duration of pretreatment from 1 to 4 h. Thus, the most active catalyst Pt/MWCNT-4 showed the highest dispersion and uniformity of Pt particles. This was favorable for the formation of hydrogenated products.

Monometallic Pt and Pd catalysts and also bimetallic catalysts containing either Pt or Pd alloyed with a transition metal (Fe, Ni, Co, Cr and Mn) supported on various supports were prepared by Liu *et al.*[49] For the preparation of MWCNT-supported catalysts, the support was purified and functionalized by nitric acid treatment, followed by washing and drying at

Table 7.13 Textural properties of nitric acid-treated MWCNTs.[48]

Sample	Surface area/ $m^2\,g^{-1}$	Pore volume/ $cm^3\,g^{-1}$	Micropore size/nm	Mesopore size/nm
MWCNT	257	0.56	0.28	5.4
MWCNT-1	293	0.94	0.26	3.5
MWCNT-2	315	0.94	0.26	3.6
MWCNT-4	254	0.80	0.26	3.7

Table 7.14 Effect of Pt/MWCNT pretreatment on furfural conversion and product selectivity.[48]

		Selectivity/%				
Catalyst	Conversion/%	FFA[a]	2-MF	THFFA[b]	DFE[c]	HP
Pt/MWCNT-1	88.1	62.4	12.7	1.3	23.7	76.4
Pt/MWCNT-2	90.7	76.8	13.2	0.7	9.3	90.7
Pt/MWCNT-4	94.4	79.0	15.3	—	5.7	94.3

[a]FFA, furfuryl alcohol; 2-MF, 2-methylfuran;
[b]THFFA, tetrahydrofurfuryl alcohol;
[c]DFE, difurfuryl ether; HP, hydrogenated products.

100 °C. In parallel, an AC as received and another treated with nitric acid (H-AC) were also used as supports for comparison, in addition to oxidic supports such as MgO, Al_2O_3, SiO_2 and ZrO_2. The monometallic catalysts were prepared by impregnation of the supports using an aqueous solution of either H_2PtCl_6 or H_2PdCl_4 followed by sonication, drying and reduction at 400 °C in H_2. The bimetallic catalysts were prepared by co-impregnation using the same Pt and Pd precursors together with the transition metals precursors. The catalyst activity and selectivity with respect to FFA were compared in the HYD of FF in ethanol (batch process, 100 °C, 3 MPa H_2, 5 h). Table 7.15[49] compares MWCNTs alone with monometallic Pt and Pd catalysts supported on MWCNTs, AC, H-AC and Al_2O_3 with corresponding bimetallic catalysts. The MWCNT support alone was inactive, giving almost 99% of an unwanted 2-furaldehyde diethyl acetal (FDA). Among the transition metals, Fe and Ni had the most beneficial effect. The superior performance of Pt–Fe and Pt–Ni supported on MWCNT was confirmed by the high conversion and selectivity with respect to FFA, giving very low yields of unwanted by-products (*e.g.* FDA and others). There is a significant difference between the selectivities of the Pt- and Pd-containing catalysts. The former is selective for FFA formation whereas Pd/MWCNT is selective for tetrahydrofurfuryl alcohol (THFFA) formation. The catalysts supported on other carbon supports and γ-Al_2O_3 were less active. The catalysts supported on oxidic supports such as MgO, ZrO_2 and TiO_2 were less active than those supported on γ-Al_2O_3.

Kang *et al.*[50] compared a commercial Pt(5%)/AC catalyst with Pt(0.47%)/MWCNT and Pt/Al_2O_3 catalysts in the HYD of 2,5-dimethylfuran (DMF) and 2-methylfuran (2MF) in a continuous system at 353–573 K and with atmospheric pressure H_2. Complete conversion of DMF over the entire temperature range was observed over the Pt/MWCNT and commercial Pt/AC catalysts. The latter gave a yield of 2-hexanol of about 80% whereas Pt/MWCNT was selective for hexane formation. During the HYD of 2MF, both catalysts exhibited similar activity and selectivity. Thus, at 573 K, at complete conversion of 2MF, a yield of 2-pentatnone of about 80% was observed. However, it should be noted that for the commercial catalysts, the amount of Pt was about 10 times that for the Pt/MWCNT catalyst. The Pt/Al_2O_3 catalyst was less reactive. Priecel *et al.*[51] prepared Ru(5%) catalysts

Table 7.15 Conversion and selectivity in the HYD of furfural over Pt- and Pd-containing catalysts supported on MWCNTs.[49]

		Selectivity[b]/%			
Catalyst[a]	Conversion/%	FFA	THFA	FDA	Others
MWCNT	24.1	—	—	98.7	1.3
Pt/MWCNT	64.5	37.6	10.7	45.9	5.8
Pt–Cr/MWCNT	85.4	69.4	2.0	26.3	2.3
Pt–Mn/MWCNT	85.9	75.7	2.0	21.3	1.0
Pt–Fe/MWCNT	95.2	91.8	2.0	5.2	1.0
Pt–Co/MWCNT	86.7	80.3	2.0	14.5	3.2
Pt–Ni/MWCNT	95.9	84.1	12.1	2.7	1.1
Pd/MWCNT	84.6	13.2	60.5	14.0	12.3
Pd–Cr/MWCNT	87.4	9.5	64.1	11.8	14.6
Pd–Mn/MWCNT	87.8	21.1	63.0	10.4	5.5
Pd–Fe/MWCNT	88.7	26.1	54.6	8.7	10.6
Pd–Co/MWCNT	86.8	12.0	70.2	11.9	5.9
Pd–Ni/MWCNT	92.7	2.1	78.2	8.9	10.8

[a]Conditions: batch process, 0.5 mL furfural in ethanol, 100 °C, 3 MPa H_2, 5 h.
[b]FFA, furfuryl alcohol; THFA, tetrahydrofurfuryl alcohol; FDA, 2-furaldehyde diethyl acetal.

supported on CNTs and ACs using an incipient wetness impregnation method with an aqueous solution of $RuCl_3{\cdot}xH_2O$. The catalysts were compared in the HYD of hydroxymethylfurfural (HMFF) to 2,5-dimethylfuran (DMF). The tests were conducted in a batch reactor at 130–200 °C and H_2 pressures of 10–40 bar. For testing, HMFF and tridecane as internal standard were dissolved in dioxane. It is well established that the reaction begins with the HYD of an aldehyde to a hydroxyl group giving an intermediate with two hydroxyl groups, followed by their hydrogenolysis to DMF. The wide ranges of temperature and H_2 pressure were used to establish the optimal conditions to obtain a high conversion of HMF with high selectivity with respect to DMF. Figure 7.5[51] shows the results obtained over Ru/AC and Ru/CNT catalysts and the commercial Ru/C catalyst. Hence the superiority of the Ru/CNT catalyst is evident. This superiority can be partly explained on the basis of the differences in textural properties. However, the unique electronic effects of the CNT support on Ru are responsible for an active site ensuring optimal adsorption of the reactant and hydrogen activation.

Wang *et al.*[52] prepared monometallic Pt and bimetallic Pt–Co catalysts supported on MWCNTs using the atomic layer deposition method. The catalysts were compared in the conversion of 5-hydroxymethylfurfural (HMFF) dissolved in 1-butanol. A high yield of 2,5-dimethylfuran (DMF) was the objective. The experiments were conducted in a batch reactor at 160 °C and 1 MPa of H_2. For comparison, Pt/CNT, Pt/SiO_2 and Pt/Al_2O_3 catalysts were also prepared and tested. In all respects, Pt/CNT outperformed the other catalysts. As the best catalyst, Pt/CNT was selected for further evaluation in combination with Co in a bimetallic Pt–Co/CNT catalyst. Complete conversion of the HMFF was observed over both Pt/CNT and Pt-Co/CNT catalysts, but the selectivity

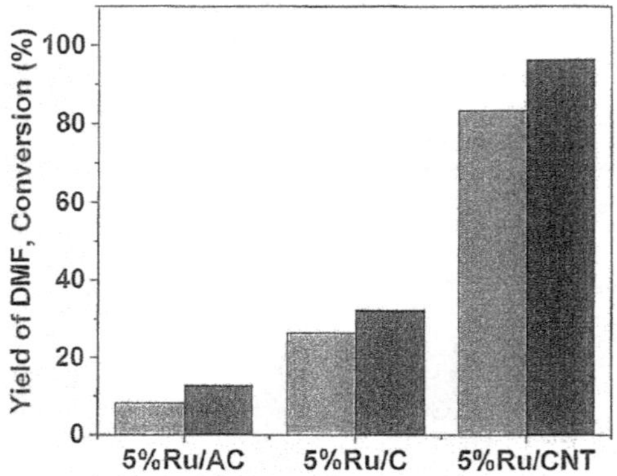

Figure 7.5 Effect of type of carbon support for Ru catalysts on the yield of DMF and turnover frequencies (TOFs) in the conversion of HMF: grey bar = DMF yield; darker bar = conversion.
Reproduced from ref. 51 with permission from American Chemical Society, Copyright 2018.

with respect to DMF over the latter catalyst reached 92% compared with less than 60% observed over the Pt/CNT catalyst.

Rarely, cyclopentanol has been a targeted product of the conversion of furfural, as reported by Mironenko *et al.*[53] They compared monometallic Pd(1.5%)/CNT and Ru(1.5%)/CNT catalysts and also a bimetallic Pd(0.85%)–Ru(0.80%)/CNT catalyst to study this reaction in a batch reactor at 473 K and 8 MPa of H_2 in the aqueous phase. The catalysts were prepared by the incipient wetness impregnation of CNTs using metal chloride precursors, followed by drying in air at 393 K and then in argon at 423 K and reduction in H_2 at 523 K. Several by-products, such as tetrahydrofurfuryl alcohol, 2-methyl-THF and cyclopentanone, were formed in addition to the targeted cyclopentanol. Complete conversion of the furfural reactant was observed over each catalyst. However, there was a significant difference in the selectivity with respect to cyclopentanol. Thus, over the Pd–Ru/CNT catalyst, the yield of cyclopentanol approached 77% after 1 h compared with 14% over the Pd/CNT catalyst after 3 h, and the selectivity of the Ru/CNT catalyst was even lower than that of the latter catalyst. Liu *et al.*[54] studied the HYD of furfural over an Ru/CNT catalyst in an autoclave at 160 °C and 1 MPa of H_2 for 5 h. The catalyst was prepared by incipient wetness impregnation of the oxidized CNTs using an aqueous solution of $RuCl_3$. Under aqueous conditions, at 99% conversion, the selectivity with respect to cyclopentanone exceeded 90%. Under the same conditions, no cyclopentanone was formed in ethanol at 57% conversion. Thus, in addition to HYD, structural rearrangement of the reactant was observed.

7.1.3.4 Levulinic Acid (LA)

According to the mechanism in Figure 5.1, levulinic acid (LA) can be formed *via* the gradual transformation of cellulose. In most studies on the conversion of LA, the main product of interest was γ-valerolactone (GVL). Based on the mechanism in Figure 5.4, GVL is obtained from LA involving both HYD and dehydration reactions. Piskun *et al.*[55] studied the HYD of LA to GVL in water over Ru(1%) catalysts supported on different oxidic supports (Al_2O_3, SiO_2, TiO_2, ZrO_2, Nb_2O_5 and zeolite β-12.5) in comparison with an Ru/CNT catalyst. The testing was conducted in a batch reactor at 90 °C and 4.5 MPa of H_2 The best performance was exhibited by the Ru/β-12.5 catalyst, giving almost complete conversion of LA with 66% selectivity with respect to GVL in 2 h. With regard to conversion and selectivity, the Ru/CNT catalyst outperformed all other Ru catalysts except the Ru/β-12.5 catalyst.

MWCNTs treated with nitric acid were used by Du *et al.*[56] as the support for the preparation of Ir/CNT catalysts. In parallel, similarly pretreated AC was also used as the support. Ir(4.5%)/CNT catalyst was prepared by the impregnation of CNTs using an aqueous solution of H_2IrCl_6, followed by washing and drying at 100 °C. In addition, the catalysts were reduced in H_2+Ar at 450 °C for 2 h. Other catalysts were prepared by a similar impregnation method. The activities and selectivities of the catalysts were compared in the conversion LA to GVL at 50 °C and 2 MPa of H_2 in a batch reactor. Other experiments were performed with an equimolar amount of formic acid at 50 °C and 4 MPa of H_2. Biomass-derived feed consisting of cellulose, starch, sucrose, glucose and/or fructose at 50 °C were also tested. The effect of the solvent on the LA conversion and selectivity with respect to GLV over an Ir/CNT catalyst is shown in Table 7.16.[56] Among the organic solvents, the best catalyst performance was observed in $CHCl_3$. Apparently, the presence of a C–O bond in other organic solvents interfered during the adsorption of LA at the active site. Such interference was not observed in water. On the basis of conversion, selectivity and yield of GVL, the Ir/CNT catalyst was superior to the Ir catalysts supported on oxidic supports and an AC support in the HYD of LA in water, as indicated by the results in Table 7.17.[56] Under aqueous conditions, the Ir/CNT catalyst was also better than the corresponding Ru, Pt and Pd catalysts.

7.1.3.5 Lignin-derived Monomers

Compared with cellulose and hemicellulose, lignin is the most stable component of lignocellulosic biomass, as indicated by a much higher resistance

Table 7.16 Effect of solvent on conversion, selectivity and yield in the HYD[a] of LA over an Ir/CNT catalyst.[56]

Solvent	Conversion/%	Selectivity/%	Yield/%
$CHCl_3$	93	98	91
Toluene	68	97	66
Methanol	20	90	18
Acetone	7	99	7
Dioxane	2	100	2
Water	100	99	99

[a]Conditions: 50 °C, 2 MPa H_2, 1 h.

Table 7.17 Effect of catalyst type on the conversion of LA to GLV[a] by HYD[b] in water.[56]

Catalyst	Conversion/%	Selectivity/%	Yield/%
Ir/CNT	100	99	99
Ir/CNT[b]	96	97	95
Ir/AC	78	98	76
Ir/CeO_2	65	99	65
Ir/TiO_2	55	99	55
Ir/Al_2O_3	52	98	51
Ir/ZnO	51	99	51
Ir/SiO_2	38	97	37
Ir/MgO	20	99	20
Ru/CNT	65	100	65
Ru/Al_2O_3	22	98	22
Pt/CNT	53	97	51
Pd/CNT	19	97	18
Pd/AC	7	100	7

[a]Conditions: 50 °C, 2 MPa H_2, 1 h.
[b]After four cycles.

to depolymerization. The monomeric products of lignin depolymerization such as phenol, alkylphenols, vanillin, guaiacol and eugenol (Figure 5.5) have been used as reactants to evaluate the activity of catalysts supported on CNTs. Generally, the objective has been conversion to hydrocarbons. Therefore, HDO was a key reaction in the overall conversion. This suggests that the experimental conditions approached those employed during HPR.

Monometallic Ru(4%) and Re(3.6%) supported on CNTs, ACs and CBs were prepared by Jung *et al.*[57] using the incipient wetness impregnation method. An aqueous solution of $RuCl_3$ and NH_4ReO_4 was used, respectively. In addition, bimetallic Ru–Re catalysts were prepared by successive impregnation of the Ru catalysts with an aqueous solution of NH_4ReO_4. The catalysts were dried at 105 °C and reduced at 250 °C in a flow of H_2(5%)–N_2. Ru catalysts promoted with Mo, W and Fe were prepared by incipient wetness impregnation of the Ru catalysts with aqueous solutions of $H_{24}Mo_7N_6O_{24}$, $H_{42}N_{10}O_{42}W_{12}$ and $Fe(NO_3)_3 \cdot 9H_2O$, respectively. The activities and selectivities of the catalysts were compared in the conversion of guaiacol in a batch reactor at 200 °C and 2 MPs of H_2. Almost complete conversion was achieved over Ru–Re/CNT and Ru–Re/CB catalysts compared with less than 60% achieved over the Ru–Re/AC catalyst. Over Ru–Re/CNT, Ru–Re/CB and Ru-Re/AC catalysts, the selectivity with respect to cyclohexane was about 40, 34 and 15%, respectively. Monometallic Ru and Re catalysts supported on these carbon supports were also prepared and tested. Under the same conditions, little activity of the Re/CNT catalyst compared with high activity of the Ru/CNT catalyst was observed. The activity of Re/CNT alone was very low, but complete conversion was achieved using a physical mixture of Ru/CNT and Re/CNT catalysts. In combination with W, Mo and Fe, Ru/CNT catalyst was less active.

Chen *et al.*[58] studied the HDO of eugenol to cycloalkanes in a biphasic water–*n*-dodecane medium over Ru catalysts supported on CNTs. The Ru/CNT catalysts were prepared by incipient wetness impregnation using an aqueous solution of $RuCl_3 \cdot 3H_2O$, which was then added to a slurry containing CNTs. The mixture was stirred at room temperature for 12 h. After separation, the solid was dried at 120 °C and reduced in $H_2 + N_2$ at 250 °C for 4 h. Using a similar procedure, Ru catalysts supported on AC, ZrO_2 and CeO_2 were prepared for comparison. The experiments were conducted in a batch reactor at 220 °C and 5 MPa of H_2. Typically, 1 mmol of eugenol was mixed with a mixture of 6 mL each of *n*-dodecane and water. A high selectivity with respect to hydrocarbons was the objective of the tests. In every run, complete conversion of eugenol was observed. Table 7.18[58] shows the superiority of the Ru/CNT catalyst relative to Ru/ZrO_2 and Ru/CeO_2 catalysts, particularly in the *n*-dodecane–water medium. Under the same conditions, the performances of Pt/AC, Pd/AC and Rh/AC catalysts were also evaluated. Although complete conversion of eugenol was achieved, the HDO selectivity of these catalysts was low, as indicated by the predominance of O-containing products (*e.g.* products C and D in Table 7.18). Additional tests were performed using different solvents, *i.e.* methyl isobutyl ketone, γ-valerolactone, cyclohexane, *n*-octane, *n*-tetradecane and *n*-dodecane. All hydrocarbon–water solvents gave a higher yield of propylcyclohexane than in a pure water solvent, whereas the O-containing solvents gave a lower yield than water. In addition to eugenol, more than a dozen lignin-derived monomers (*e.g.* phenols, anisole, guaiacol, catechol) were tested over the Ru/CNT catalyst. In most cases, a selectivity with respect to hydrocarbons of >90% was obtained.

The conversion of vanillin (4-hydroxy-3-methoxybenzaldehyde) was performed in an autoclave at 100 °C and 1 MPa of H_2 for 3 h with continuous stirring.[59] For the tests, a slurry of Ru/CNT catalyst, vanillin, decalin and water was used. Fluorescence images revealed that the Ru/CNT catalyst particles were located on the surface of the emulsion droplets. Apparently,

Table 7.18 Product distribution in the HYD[a] of eugenol.[58]

		Product distribution[c]/%			
Catalyst	Solvent[b]	A	B	C	D
Ru/AC	S1	59.5	3.1	0.3	—
	S2	21.5	26.2	35.2	—
Ru/CNT	S1	56.5	2.1	0.4	—
	S2	94	1.0	—	—
Ru/ZrO_2	S1	29.2	2.4	24.7	0.1
	S2	45.7	5.5	40.0	0.2
Ru/CeO_2	S1	7.4	12.5	47.0	0.2
	S2	12.7	60.5	25.6	0.2

[a]Conditions: 220 °C, 5 MPa H_2, 4 h.
[b]Solvent S1 = water; solvent S2 = water–*n*-dodecane.
[c]A, propylcyclohexane; B, 4-propylcyclohexanol; C, 2-methoxy-4-propylcyclohexanol; D, 2-methoxy-4-propylphenol.

the Ru/CNT acted as an emulsifying agent and catalyst. The effect of the decalin-to-water ratio on the catalyst activity and selectivity is shown in Table 7.19.[59] The objective of the study was the conversion of vanillin to vanillyl alcohol rather than complete HDO to hydrocarbons.

Wang *et al.*[60] showed that the dehydration of vanillyl alcohol can be achieved under very mild conditions, *i.e.* room temperature. They prepared Pd(5%) catalysts supported on pristine CNTs, nitric acid-oxidized CNTs, ACs and an ultra-dispersed diamond powder. The highest activity was exhibited by the catalyst supported on oxidized CNTs. This was attributed to the presence of acidity supplied by the carboxylic groups on the oxidized CNTs. Carbon solids such as MWCNTs, carbon aerogel (CARF), CB (Vulcan carbon), AC and graphite were compared as supports of Ru catalysts that were used for the HDO of guaiacol in an autoclave at 270–300 °C and 4 MPa of H_2 for 1 h.[61] The activities of the catalysts, in terms of the guaiacol conversion and overall HDO, are compared in Figure 7.6. The activity order Ru/MWCNT > Ru/AC > Ru/CARF > Ru/Vulcan carbon > Ru/graphite was established. The highest activity coincided with the highest ratio of

Table 7.19 Effect of decalin-to-water ratio on the activity and selectivity of Ru/CNT catalyst[a] in the conversion of vanillin.[59]

Decalin-to-water ratio/mL/mL	Conversion/%	Selectivity/% Vanillyl alcohol	*p*-Cresol
40/0	24	89	11
30/10	86	74	26
20/20	97	74	26
10/30	99	76	24
0/40	95	87	13

[a]Conditions: 5.9 mmol vanillin, Ru/CNTs (Ru: 0.2 mol%), 100 °C, 1 MPa H_2, 3 h.

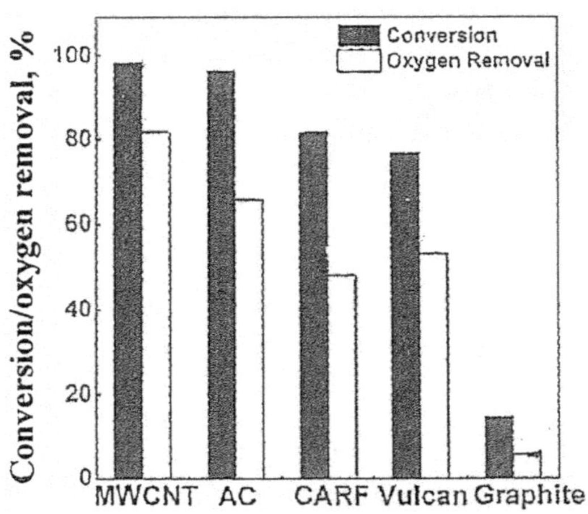

Figure 7.6 Effect of type of carbon support for Ru catalysts on HDO rate and conversion of guaiacol.[61]

mesoporosity to microporosity of the Ru/MWCNT catalyst. The high activity of the Ru/MWCNT catalyst was further confirmed using a mixture of model compounds such as eugenol, cerulignol and vanillin. These reactants were completely converted to the corresponding alcohols and hydrocarbons. The high activity of the Ru/MWCNT catalyst was confirmed in an additional test conducted under aqueous conditions. In this case, bio-oil obtained from the pyrolysis of oak chips was used as the feed, *i.e.* 1.276 g of bio-oil, 40 mL of water and 0.6 g of catalyst at 300 °C. The originally brown mixture changed into a colorless liquid after the test.

In a study by Yohe *et al.*,[62] the HDO of dihydroeugenol over Pt(5%)–Mo(2.5%)/MWCNT catalysts was compared with that over the corresponding monometallic catalysts. The reaction intermediates (4-propylphenol and others) were also included. Testing was conducted in a continuous system at 573 K and ~2 MPa. Under these conditions, complete conversion to hydrocarbons over the bimetallic catalyst was achieved.

Ohta *et al.*[63] determined the activity of Pt catalysts supported either on ACs or on MWCNTs in the HDO of 2-propylphenol in water (573 K, 4 MPa of H_2, 6 h) in an autoclave. Complete conversion of the reactant over both catalysts was achieved with 97 and 92% selectivity with respect to propylcyclohexane, respectively. Li *et al.*[64] used the HYD of phenols as a model reaction to test the difference between the activities of Pt nanoparticles located in the channels (Pt-in/CNT) and those on the exterior (Pt-out/CNT) of CNTs. For the preparation of the former catalyst, CNTs were treated with nitric acid at 398 K before impregnation with an ethanolic solution of H_2PtCl_6. For the Pt-out/CNT catalyst, CNTs were impregnated with ethanol first, followed by the addition of an aqueous solution of H_2PtCl_6. After drying, the samples were reduced in H_2 at 300 °C for 3 h. In parallel, a Pt-imp/CNT catalyst was prepared by incipient wetness impregnation. The activities and selectivities of these catalysts are compared in Table 7.20.[64] Superior activity of the Pt-in/CNT catalyst is evident from the high conversion and reaction rate.

Pd and Ni catalysts supported either on CNTs or ACs were evaluated by Xiang *et al.*[65] in the controlled HYD of phenol to cyclohexanone and cyclohexanol. The experiments were conducted in a continuous fixed-bed reactor at 493 K, 3.5 MPa and LHSV 3.87 h^{-1} with the phenol reactant dissolved in aqueous methanol. In parallel, the HYD of phenol was carried out simultaneously with the aqueous phase reforming of methanol as the source of

Table 7.20 Conversion and selectivity in the HYD[a] of phenol.[64]

Catalyst	Conversion/%	Selectivity/%		Rate/mol phenol (mol Pt)$^{-1}$ s^{-1}
		Cyclohexanone	Cyclohexanol	
CNT	<1	—	—	—
Pt-in/CNT	97.3	77.5	22.5	0.934
Pt-out/CNT	1.6	72.3	27.7	0.111
Pt-imp/CNT	33.5	75.3	24.7	0.324

[a]Conditions: 20 mg catalyst, 0.5 g phenol, 0.5 MPa H_2, 323 K, 30 min.

Table 7.21 Activity and selectivity of Pd/AC and Pd/CNT catalysts in the HYD of phenol.[65]

Catalyst	Temperature/K	H_2 pressure/MPa	Phenol conversion/%	TOF/h^{-1}	Selectivity/% Cyclohexanol	Cyclohexanone
Pd/AC	493	2	71.0	15.3	73.2	26.8
Pd/CNT	493	2	99.7	39.7	64.2	34.8
Pd/CNT	423	1	87.8	35.0	94.7	0.3

in situ-generated hydrogen. In this case, the reaction stream comprised phenol mixed with methanol and water while other conditions remained the same. It was evident that the Pd and Ni catalysts supported on CNTs exhibited much higher activity than those supported on ACs. This is shown for the Pd/CNT catalyst in Table 7.21.[65] As expected, the introduction of O-containing groups into CNTs *via* treatment with nitric acid improved the hydrophilicity and increased the selectivity with respect to cyclohexanone but the overall conversion of phenol decreased. Pd/CNT catalysts were always more active than Ni/CNT catalysts. In a study by Zhang *et al.*,[66] a gradual deactivation of the Pd/CNT catalyst during the HYD of phenol was observed. This was caused by the agglomeration of Pd particles. Catalyst reactivation *via* particle redispersion could be achieved by treatment with HNO_3 vapor. Apparently, nitrogen oxides formed by HNO_3 decomposition played a key role during redispersion.

7.1.4 Conversion of Glycerides and Fatty Acids

In studies over noble metals catalysts supported on CNTs, reactants derived from vegetable oils and algae biomass have attracted less attention than lignocellulosic sources of reactants in spite of the high stability of carbon-supported catalysts in such applications. Pt/MWCNT and also Pt/AC and Ru/AC catalysts, all containing 5 wt% of the active metal, were compared in the decarboxylation of fatty acids present in microalgae biomass (*e.g.* stearic, palmitic, lauric, myristic, arachidic and behenic acids) with the aim of producing hydrocarbons for biodiesel.[67] Decarboxylation was performed in an autoclave in an aqueous medium at 330 °C. Under the same conditions, the Pt/MWCNT catalyst outperformed the other catalysts. Over this catalyst, at almost complete conversion, the selectivity for decarboxylation approached 97%.

Nitrogen-doped CNTs (NCNTs) prepared by the pyrolysis of CNTs with phenanthroline were used as supports for Ru catalysts in the conversion of stearic acid and crude vegetable oil (batch reactor, 220 °C, 4 MPa of H_2).[68] Little conversion was observed over CNTs and NCNTs without active metal being present. Over Ru/CNT and Ru/NCNT, the total yield of hydrocarbons from the conversion of stearic acid approached 69 and 95%, respectively. For comparison, Ru/TiO_2, Ru/SiO_2 and Ru/Al_2O_3 catalysts were also used. Among all the catalysts tested, the Ru/NCNT catalyst exhibited superior activity and selectivity with respect to C_{17} and C_{18} hydrocarbons during the conversion of both stearic acid and crude vegetable oil.

Studies in which carbon nanosupports were compared with oxidic supports under similar conditions are of importance. For example, Wang *et al.*[69] compared MWCNTs with ACs and also TiO_2 and ZrO_2 as supports for Pt catalysts used in the HYD of feeds of vegetable oil origin (batch process, 170 °C, 3 bar of H_2, 4 h). For the experiments, the soybean oil feed was slurried with catalyst and used directly. The objective was to reduce the formation of *trans* fatty acids and saturated acids in soybean oil. It was confirmed that the Pt/ZrO_2 catalyst exhibited the highest activity and selectivity.

7.1.5 Other Oxygenates

Xiang *et al.*[70] compared Pd/CNT with Pd/AC catalysts in the HYD of acetophenone. Significant differences in the selectivities of the catalysts were noted. For example, the yield of phenylethanol approached 95% over the Pd/CNT catalyst compared with about 48% over the Pd/AC catalyst. The tests were conducted at 333 K and atmospheric pressure H_2 in a continuous system with the acetophenone reactant dissolved in ethanol. The difference in selectivity was attributed to different adsorption modes of the reactant on the catalyst active site.

Zhou *et al.*[71] studied the effect of pretreatment of CNTs on the performance of Pd/CNT catalysts in the HYD of benzaldehyde to benzyl alcohol. Before the catalyst preparation, the CNTs were treated with nitric acid at 130 °C for 12 h. To introduce nitrogen, the oxidized CNTs (OCNTs) were treated with ammonia at 200, 400 and 600 °C. The OCNT and NCNT supports were used for the preparation of Pd (~1 wt%) catalysts by incipient wetness impregnation using an aqueous solution of H_2PdCl_4. Before the experiments, the catalysts were suspended in ethanol in a batch reactor and reduced in a flow of H_2 at 45 °C for 1 h. The reaction at 45 °C began after the addition of benzaldehyde. Figure 7.7[71] shows that the selectivity approached 100% before the overall conversion reached 70%, when the selectivity began to decrease. Above 70% conversion, the most pronounced decrease in selectivity was observed over the Pd/OCNT catalyst and the Pd/400NCNT catalyst exhibited the best performance. This was attributed to the electron-donating ability of the doped N atoms, which enhanced the desorption of benzyl alcohol from active sites.

Valuable dimethyl cyclohexanedicarboxylates (DMCDs) are prepared by the HYD of dimethyl terephthalate (DMT). Monometallic Ni and Ru and bimetallic Ru–Ni catalysts supported on CNTs were tested for such an application by Huang *et al.*[72] First, the CNTs were functionalized by treatment with nitric acid before being used as the support for wet impregnation with a solution of Ni chloride and Ru chloride in deionized water. For the preparation of Ru–Ni/CNT catalyst, the CNT support was first impregnated with Ni, followed by impregnation with Ru. The corresponding catalysts supported on AC were also included. The testing was conducted in a batch reactor at 423 K, an initial reaction pressure of 5 MPa and a reaction time of

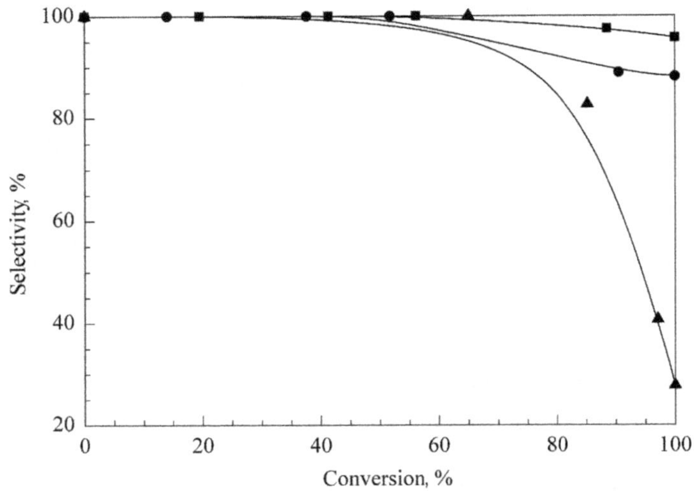

Figure 7.7 Effect of pretreatment of CNTs on the selectivity of Pd catalysts in benzaldehyde conversion: ■, Pd/400NCNT; ●, Pd/CNT; ▲, PdO/CNT.
Reproduced from ref. 71 with permission from Elsevier, Copyright 2012.

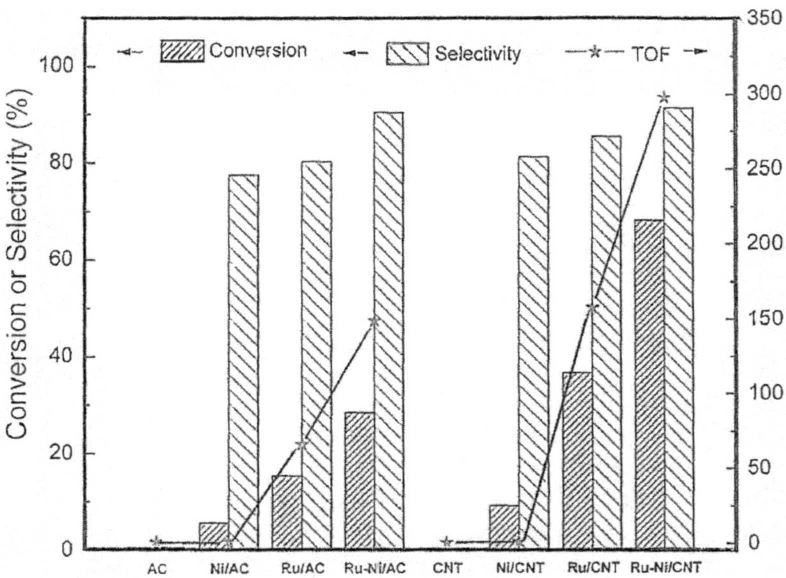

Figure 7.8 Effect of AC and CNTs on catalyst activity, selectivity and TOF in the conversion of dimethyl terephthalate.
Reproduced from ref. 72 with permission from American Chemical Society, Copyright 2014.

1 h using the reactant dissolved in ethyl acetate. In terms of conversion, selectivity and TOF, the superiority of the Ru–Ni/CNT catalyst is confirmed by the results in Figure 7.8.[72]

Tourani *et al.*[73] used rGO, MWCNTs and rGO–CNT composite as supports for the preparation of Pd nanocatalysts. These catalysts were compared with a Pd/AC catalyst in the hydropurification of crude terephthalic acid (CTA). In spite of the unusual feed, the conditions employed approach those observed in other studies. Moreover, in a rare case, Pd catalysts supported on four different carbon supports were evaluated under identical conditions, *i.e.* batch reactor, 290 °C, 1 MPa, 1 h. The composition of the feed used for testing approached that in an industrial process, *i.e.* CTA (120 g) containing ~2100 ppm of 4-carboxybenzaldehyde (4-CBA), 234 ppm of *p*-toluic acid (TA) and 50 ppm of benzaldehyde (BA) as impurities in 280 mL of water slurried with 1.3 g of catalyst. The catalysts were prepared by wet impregnation aided by sonification. $PdCl_2$ was used as the Pd precursor. The rarely used rGO–CNT composite was prepared by the addition of purified MWCNTs to a homogeneous aqueous dispersion of GO in a 1 : 1 weight ratio with sonication for approximately 30 min. The former was bonded to the surface of GO *via* strong π–π interactions. The mixture was treated hydrothermally at 180 °C for 12 h, then cooled to room temperature. In terms of removal of 4-CBA and BA impurities, the Pd/rGO–CNT catalyst exhibited the highest activity. Also, it was more active than a commercial Pd/AC catalyst. In another study by Tourani *et al.*,[74] Pd and Pd–Sn both supported on functionalized MWCNTs were compared. Over the mono- and bimetallic catalysts, 99% removal of the CBA impurity was achieved. However, the Pd–Sn/MWCNT catalyst had a lower activity for the HYD of TA. The commercial Pd(0.5%)/AC catalyst was less active. The performance of the Pd/MWCNT catalyst could be further improved by optimizing the catalyst preparation procedure.[75] In this regard, the focus was on parameters such as the type of Pd precursor, Pd loading and reduction temperature.

7.1.6 Summary of Results on Conversion of Lignocellulosic Reactants and Fatty Acids

With respect to catalysts containing noble metals supported on CNTs, reactants derived from cellulose and hemicellulose and also monomeric reactants obtained from lignin have attracted much more attention than those derived from vegetable oils and algae biomass and other oxygenates. According to Table 7.22, depolymerization of the biomass components, followed by partial conversion of the primary products, has commonly been studied. For example, the production of sorbitol *via* either the direct conversion of cellulose or gradual conversion of cellulose through glucose has frequently been investigated. In this case, hydrolysis reactions played a dominant role in addition to the HYD of the aldehyde group of glucose. A superior performance of catalysts supported on CNTs compared with those supported on ACs and CBs and also oxidic supports (*e.g.* γ-Al_2O_3, SiO_2, MgO and CeO_2) in these reactions was confirmed.[39] Moreover, little loss of activity of the former catalysts was observed during several repeated utilization–regeneration cycles.

Table 7.22 Conversion of lignocellulosic reactants in batch systems.[a]

Catalyst	Reaction	Conditions[b]	Ref.
Partial conversion			
1. Ru/CNT; Ru/AC; Ru/graphite; Ru/CB; Ru/Al_2O_3; Ru/HY	CEL → SOR Xylan → xylitol	448 K; 5 MPa; aq.	36, 37
2. Ru/CNT; Pt/CNT; Rh/CNT; Co/CNT; Ni/CNT; Ir/CNT	CEL → SOR	458 K; 5 MPa; aq.	39
3. Ru/CNT; Ru/Al_2O_3; Ru/SiO_2; Ru/MgO; Ru/CeO_2	CEL → SOR	458 K; 5 MPa; aq.	39*a*
4. Ru/CNT	GLU → SOR	403 K; 2 MPa; aq.	40
5. Ru/CNT	CEL and GLU → SOR	478 K; 5 MPa; aq.	41
6. Ni/AC; Ru–Ni/CNT; Ru–Ni/AC	CEL → SOR	478 K; 5 MPa; aq.	42
7. Ru/CNT; Ru/Al_2O_3; Ru/SiO_2	GLU → SOR	393 K; 4 MPa; aq.	43
8. Ru/CNT	SOR → glycols	483 K; 6 MPa; aq.	44
9. Ru/CNT; Pd/CNT, Pt/CNT, Rh/CNT; Au/CNT	XYL → FFA, xylitol	403 K; 3 MPa; aq.	46
10. Pt/CNT; Pt/AC; Ru/AC; RANEY® Ni	XYL → FFA, xylitol	393 K; 5.5 MPa; aq.	10, 27, 28
11. Ru/CNT; Ru/AC; Pd/CNT; Pd/AC	FF → FFA	363 K; 2 MPa; aq.	47
12. Pt/CNT	FF → FFA	423 K; 2 MPa; propanol	48
13. Pt/CNT; Pd/CNT; Pt/AC; Pd/AC; alloys of Fe, Ni, Co, Cr and Mn	FF → FFA	373 K; 3 MPa; ethanol	49
14. Pt/CNT; Pt/AC; Pt/Al_2O_3	DMF → 2MF	353–573 K; 1 atm H_2; vapor phase	50[c]
15. Ru/CNT; Ru/AC	HMF → DMF	403–473; 1–5 MPa; dioxane	51
16. Pt/CNT; Pt–Co/CNT	HMF → DMF	433 K; 1 MPa; butanol	52
17. Pd/CNT; Ru/CNT; Pd–Ru/CNT	FF → cyclopentanol	473 K; 8 MPa; aq.	53
18. Ru/CNT	FF → cyclopentanone	433 K; 1 MPa; aq.	54
19. Ru/CNT; Ru supported on Al_2O_3, SiO_2, TiO_2, ZrO_2, Nb_2O_5 and zeolite	LA → GVL	363 K; 4.5 MPa; aq.	55
20. Ir/CNT; Ru/CNT; Pt/CNT; Pd/CNT	LA → GVL	323 K; 2 MPa; aq.	56
21. Ru/CNT	VAN → VANALC	373 K; 1 MPa; decalin	59

Conversion to hydrocarbons			
22. Ru–Mo/CNT; Ru–Mo/AC	SOR → hydrocarbons	573 K; 4 MPa; aq.	45[c]
23. Pt/CNT	FF → hydrocarbons	423 K; 2 MPa	48
24. Ru/CNT; Ru/AC; Ru/CB; Ru/graphite	Guaiacol → hydrocarbons	573 K; 4 MPa; aq.	61
25. Ru/CNT; Ru/CB; Ru/AC; Re/CNT; Re/CB; Re/AC; Ru–Re/CNT; Ru–Re/CB; Ru–Re/AC	Guaiacol → hydrocarbons	473 K; 2 MPa; *n*-heptane	57
26. Ru/CNT; Ru/AC; Pt/AC; Pd/AC; Ru/ZrO_2; Ru/CeO_2	Eugenol → hydrocarbons	493 K; 5 MPa; aq.	58
27. Pt/CNT; Mo/CNT; Pt–Mo/CNT	DiH-eugenol → hydrocarbons	573 K; 2 MPa; vapor phase	62[c]
28. Pt/CNT; Pt/AC	Propylphenol → propylcyclohexane	573 K; 4 MPa; aq.	63
29. Pt/CNT	Phenol → cyclohexane	323 K; 0.5 MPa; dichloromethane–water	64
30. Pd/CNT; Ni/CNT; Pd/AC; Ni/AC	Phenol → cyclohexanol	493 K; 3.5 MPa; aq. methanol	65
31. Pd/CNT	Phenol → hydrocarbons	353 K; 0.1 MPa; aq.	66
32. Pt/CNT; Pt/AC; Ru/AC	Fatty acids → hydrocarbons	603 K; aq.	67
33. Ru/CNT; Ru/TiO_2, Ru/SiO_2; Ru/Al_2O_3	Stearic acid → hydrocarbons	493 K; 4 MPa	68
34. Pt/CNT; Pt/AC; Pt/TiO_2; Pt/ZrO_2; RANEY® Ni	Fatty acids → hydrocarbons	443 K; 0.3 MPa	69
Other oxygenates			
35. Pd/CNT; Pd/AC	Acetophenone → phenylethanol	333 K; 1 atm H_2; ethanol	70[c]
36. Pd/CNT	Benzaldehyde → benzyl alcohol	318 K; 1 atm H_2; ethanol	71
37. Ru/CNT; Ni/CNT; Ru–Ni/CNT	DMCD → DMT	423 K; 5 MPa; ethyl acetate	72
38. Pd/rGO; Pd/CNT; Pd/AC; Pd/rGO–CNT; Pd–Sn/CNT	CTA → TA	563 K; 1 MPa H_2; aq.	73–75

[a]CEL, cellulose; SOR, sorbitol; GLU, glucose; FF, furfural; FFA, furfuryl alcohol; HMF, hydroxymethylfurfural; DMF, 2,5-dimethylfuran; 2MF, 2-methylfuran; LA, levulinic acid; GVL, γ-valerolactone; VAN, vanillin; VANALC, vanillyl alcohol; DMCD, dimethyl cyclohexanedicarboxylates; DMT, dimethyl terephthalate; CTA, crude terephthalic acid; 4-CBA, 4-carboxybenzaldehyde.
[b]Conditions: aq., aqueous; vp, vapor phase.
[c]Continuous system.

Furfural (FF) and related reactants, all derived from either cellulose or hemicellulose (Figure 5.1), are preferentially converted to high-value products such as methylfurans rather than to hydrocarbons. Less severe conditions are required for the former products. Generally, the conditions and catalysts used for the conversion of these reactants approached those employed in the conversion of CALD and citral (Table 7.8). Similarly, the conversion of LA to GVL (Figure 5.4) could be achieved under mild conditions; it involves the HYD of one double bond, followed by dehydration.

Most of the results for the conversion of reactants and feeds derived from cellulose and hemicellulose were obtained in batch reactors under aqueous conditions. A few studies were conducted with reactants dissolved in polar solvents. The use of hydrocarbon solvents was avoided. This may be attributed to the poor miscibility of highly polar reactants and feeds with hydrocarbons. The depolymerization of lignin (Figure 5.5) involved the cleavage of numerous chemical bonds before monomeric products could be obtained. Therefore, a high HPR severity must be employed to achieve the desired conversions.[76] This is also evidenced by the high consumption of hydrogen. Studies on the further conversion of depolymerization products (*e.g.* phenols, guaiacol, eugenol) shown in Figure 5.5 were focused on hydrocarbons as final products. Therefore, the severity of the conditions approached that employed in the HDO of phenols.[76,77] Similarly, fatty acids derived from vegetable oils and algae biomass were converted to diesel fuel hydrocarbons. Most of testing was carried out in an aqueous medium. Therefore, the separation of hydrocarbons from the reaction mixture is fairly simple.

It is evident from Table 7.22 that whenever hydrocarbons are the final products, the conditions approached those employed during HPR. The preceding discussion indicated that in all studies in which catalysts supported on amorphous carbon supports (*e.g.* CBs and ACs) and those supported on oxidic supports used for comparison, the activities and selectivities of noble metal catalysts supported on CNTs were much higher. Several studies also confirmed much higher stabilities of CNT-supported catalysts, particularly under aqueous conditions. Moreover, high activity and selectivity of the catalysts were maintained during several repeated cycles.

Among the various catalysts, Pt/CNT, Pd/CNT and Ru/CNT have been the most frequently used. To a lesser extent, Rh/CNT and Ir/CNT catalysts were also tested. The combinations of two metals, *i.e.* either two noble metals or one noble metal with one transition metal (*e.g.* Co, Ni, Fe, Cu), supported on CNTs resulted in enhanced catalyst performance and stability. The selection of the best catalyst for a particular case requires extensive testing under a wide range of experimental conditions.

7.1.7 Conversion of Hydrocarbons

Of particular interest are hydrocarbons derived from conventional and nonconventional petroleum. In this case, the primary objective has been the removal of S and N heteroatoms, hydrodearomatization (HDAr) and hydroisomerization (HIS) to satisfy specifications for commercial fuels and lubricants.[78] A final polish of petroleum products may require the removal of unsaturated hydrocarbons to ensure the stability of fuels and lubricants. There are industrially important gaseous streams (*e.g.* pyrolysis products) that require mild HYD to remove undesirable by-products, *e.g.* even traces of acetylene and phenylacetylene must be removed before ethylene and styrene monomer, respectively, can be used for polymer production. In relevant studies, fairly mild conditions have been employed. The removal of aromatics *via* HDAr may be required to satisfy regulatory levels in transportation fuels. Catalysts comprising noble metals supported on CNTs may be suitable for these applications.

7.1.7.1 Aromatics

Yoon *et al.*[79] studied monometallic Pt, Pd and Rh and bimetallic Pd–Rh catalysts, all supported on CNTs, in the HYD of benzene, toluene and phenylcyclohexene at room temperature and 1 MPa of H_2 in a batch system. Figure 7.9[79] shows that the bimetallic Pd–Rh catalyst was the most active whereas Pd/CNT exhibited little activity. Au/CNT was also tested and exhibited little activity, similarly to the CNTs tested without active metals. The HYD of toluene to methylcyclohexane was performed in similar way. Again, the Pd–Rh catalyst was the most active. In the case of phenylcyclohexene,

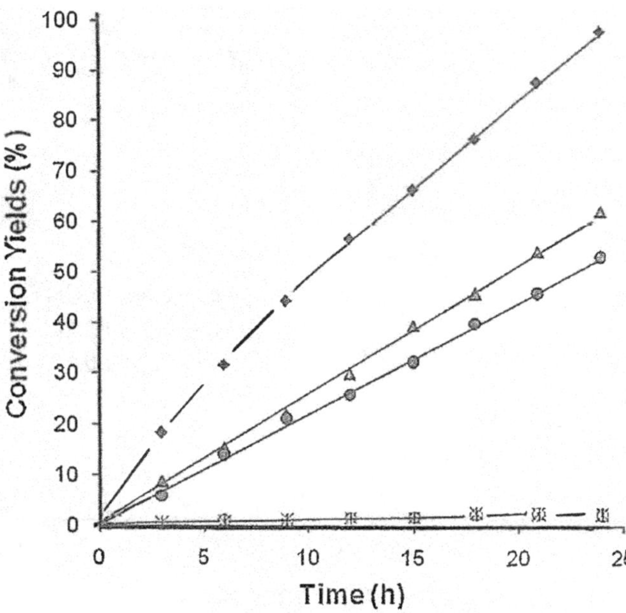

Figure 7.9 Effect of catalyst type on the conversion of benzene to cyclohexene: ♦, Pd–Rh/CNT; ▲, Pt/CNT; ●, Rh/CNT; ☼, Pd/CNT.
Reproduced from ref. 79 with permission from American Chemical Society, Copyright 2009.

HYD of the double bond to yield phenylcyclohexane was observed whereas HYD of the aromatic ring was not evident. Also for this reaction, the Pd–Rh/CNT catalyst was the most active. Successful deposition of Ru nanoparticles on CNTs using $RuCl_2 \cdot H_2O$ as precursor was achieved in supercritical water (at 400 °C) by Sun *et al.*[80] The Ru/CNT catalyst was used in the HYD of benzene at 308 K and 3 MPa of H_2 in a batch reactor. After 2.1 h, complete HYD of benzene to cyclohexane was achieved. Over a bimetallic Pt–Rh/CNT catalyst, the HYD of benzene, toluene and polycyclic aromatics to saturated hydrocarbons could be achieved at room temperature.[81] The catalysts could be recovered and reused six times without losing activity. CNTs prepared by the catalytic decomposition of CH_4 *via* the CVD method and purified by treatment with boiling nitric acid, washing with deionized water and drying at 383 K in air were used as the support for a Pt(1%)/CNT catalyst. This catalyst was compared with a reference Pt/AC catalyst.[82] To test the catalysts, the HYD–HDAr of toluene or tetralin was conducted in a continuous fixed-bed system at 373 K, 0.4 MPa and GHSV = 120 L h^{-1} $g_{cat.}^{-1}$ for toluene and 373 K, 0.16 MPa and GHSV = 90 L h^{-1} $g_{cat.}^{-1}$ for tetralin. In all respects, the Pt/CNT catalyst exhibited much higher activity than the Pt/AC catalyst. Thus, as Figures 7.10 and 7.11[82] show, the amount of Pt on AC had to be more than doubled to approach the activity of the Pt/CNT catalyst in the HYD of toluene and tetralin, respectively (Figures 7.12 and 7.13).[83]

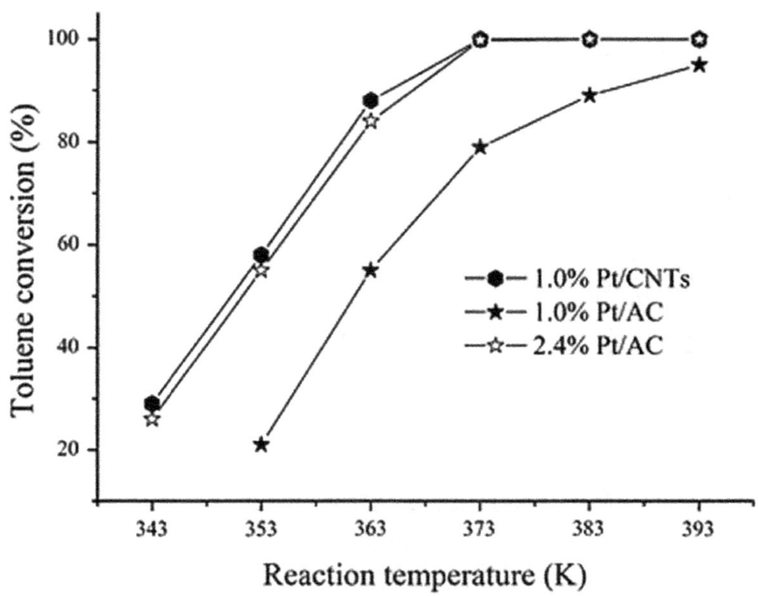

Figure 7.10 Effect of AC and CNT supports on the activity of Pt catalysts in the HYD of toluene.
Reproduced from ref. 82 with permission from Springer Nature, Copyright 2010.

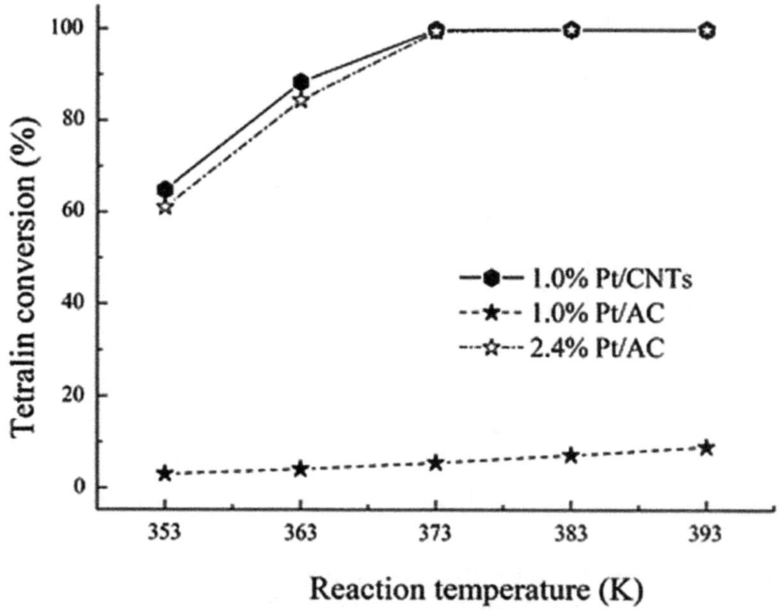

Figure 7.11 Effect of AC and CNT supports on the activity of Pt catalysts in the HYD of tetralin.
Reproduced from ref. 82 with permission from Springer Nature, Copyright 2010.

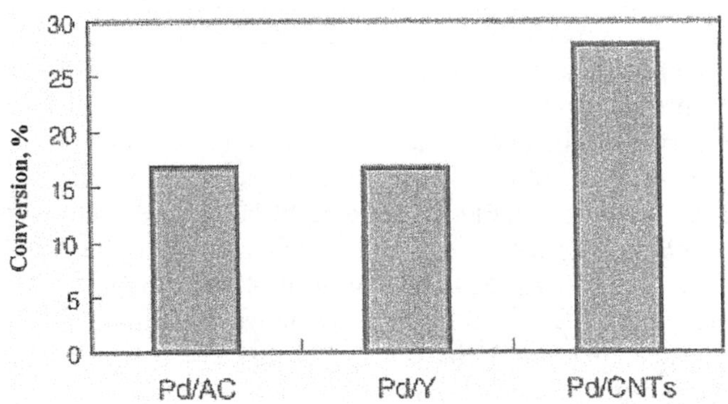

Figure 7.12 Effect of AC, zeolite Y and CNT supports on the activity of Pd catalysts in the conversion of benzene.
Reproduced from ref. 83 with permission from Elsevier, Copyright 2004.

In a comprehensive study on the HYD of aromatics, Pawelec *et al.*[84] used MWCNTs (10–30 nm diameter, 1–10 μm length), an amorphous silica-alumina (ASA) and silica-delaminated zirconium phosphate (ZrPSi) as supports for bimetallic Pt–Pd catalysts with the aim of evaluating the resistance of

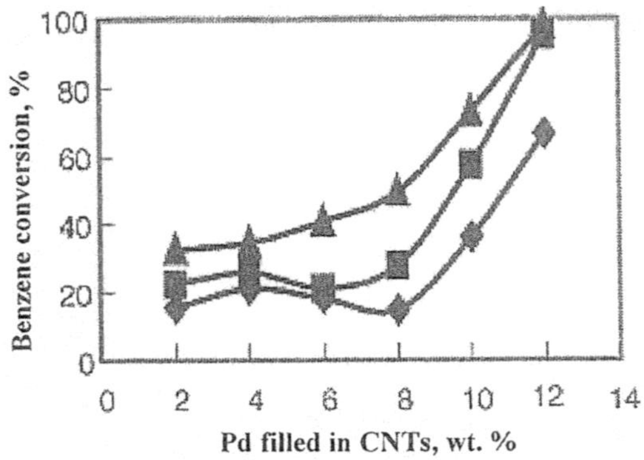

Figure 7.13 Effect of amount of Pd filled in CNTs and reaction time on the HYD of benzene: ▲, 5 h; ■, 4 h; ♦, 3 h.
Reproduced from ref. 83 with permission from Elsevier, Copyright 2004.

catalysts to poisoning by sulfur. The CNT-supported catalysts were prepared by simultaneous impregnation of the support with metal precursors such as H_2PtCl_6 and H_2PdCl_4 in aqueous solution. Prior to impregnation, the CNTs were functionalized by treatment with nitric acid. Pt–Pd/ZrPSi catalyst was prepared by consecutive wet impregnation using first an aqueous solution of $PdCl_2$ followed by impregnation with an aqueous solution of H_2PtCl_6. Pt–Pd/ASA catalyst was prepared by the adsorption–impregnation method. Prior to testing, all catalysts were reduced at 573 K for 3.5 h in a flow of a 1 : 5 v/v H_2–N_2 mixture. The performances of the catalysts were compared in the HYD of a model feed such as a mixture containing 19.1% of toluene, 2.1% of naphthalene and 100 ppm of S as dibenzothiophene (DBT) in hexadecane. The testing was carried out in a continuous system at 498 K and a total pressure of 5 MPa. For the HYD of toluene, the following order of the initial and steady-state activities was established: Pt–Pd/MWCNT > Pt–Pd/ASA > Pt–Pd/ZrPSi. However, based on Figure 7.14,[84] the performance of the Pd–Pt/MWCNT catalyst was significantly better than that of the other catalysts. This order was established in spite of the fact that the most active Pt–Pd/MWCNT catalyst had the lowest surface area and active metal loading. This confirmed the importance of electronic effects of CNT supports on active metals. The corresponding monometallic catalysts were not part of the study.

The activities of Pt/CNT, Pt/AC and Pt/Al_2O_3 catalysts were compared in the HYD and/or HDAr of toluene in a batch reactor at 343–413 K and 0.4 MPa of H_2.[85] Among the supports tested, the CNTs had the most beneficial effect because they facilitated the efficient dispersion of Pt nanoparticles. Moreover, the Pt/CNT catalyst could be reduced at much lower temperatures than the Pt/AC and Pt/Al_2O_3 catalysts. Consequently, a higher concentration of the catalytically active Pt^0 metal on the CNT support could be established. The effect of the Pt content of the Pt/CNT catalyst on the HYD activity is shown in

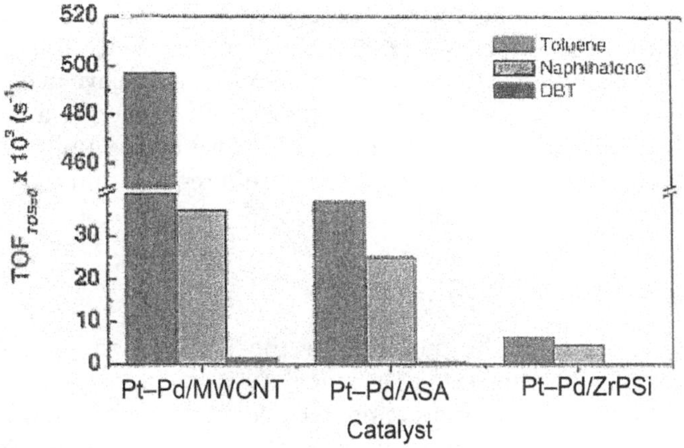

Figure 7.14 Effect CNTs, ASA and ZrPSi supports on the HYD activity of Pt–Pd catalysts.
Reproduced from ref. 84 with permission from Elsevier, Copyright 2006.

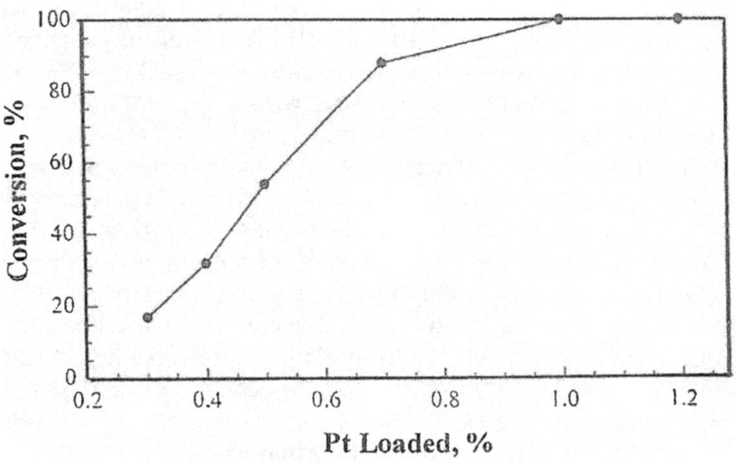

Figure 7.15 Effect of Pt loading on toluene conversion.[85]
Reproduced from ref. 49 with permission from Elsevier, Copyright 2018.

Figure 7.15.[85] It appears that little activity change is observed on increasing the Pt content above 1 wt%.

The biosynthesis-aided preparation of Ru/CNT catalysts was reported by Ma *et al.*[86] A biomass extract (*Cacumen platycladi*) was used as the reducing agent. Also, this agent ensured immobilization of Ru nanoparticles on CNTs. For the catalyst preparation, dried CNTs were mixed with an aqueous solution of $RuCl_3$ with continuous stirring, then the biomass extract was added to the mixture. The products were filtered, washed, vacuum dried at 60 °C and thermally treated under N_2 at 500 °C for 3 h. The Ru/CNT catalysts were tested in the HYD of benzene to cyclohexane in an autoclave at 353 K and 4 MPa of H_2. In parallel, commercial catalysts such as Ru/AC and Ru/Al_2O_3 were tested to compare their activities with those of the Ru/CNT catalysts from the biosynthesis-aided preparation. In this regard, remarkable activity of the latter catalyst was noted.

7.1.7.2 Unsaturated Hydrocarbons

MWCNTs were used as the support for the preparation of a Pd(0.17%) catalyst to be used for the HYD of acetylene.[87] The support was preoxidized in boiling H_2O_2 prior to impregnation with $Pd(NO_3)_2$ solution. After drying, the catalyst was calcined in air at 500 °C for 0.5 h. Another Pd(0.5%) catalyst was prepared using $[Pd(NH_3)_4]Cl_2{\cdot}H_2O$ as Pd precursor. After drying, this catalyst was reduced in H_2 at 150 °C for 2 h. The difference in the treatments of the catalysts led to different average particle sizes, *i.e.* 34 and 6 nm for Pd(0.17%) and Pd(0.5%), respectively. The activities and selectivities with respect to ethylene of the catalysts were determined in a continuous fixed-bed reactor in a flow of H_2. In the temperature range 80–200 °C, the activities of the two catalysts, expressed as the overall conversion of acetylene, were similar. However, there was a significant difference in the selectivity with respect to ethylene, the Pd(0.17%)/MWCNT with a large size of Pd particles being much more selective than the Pd(0.5%)/MWCNT catalyst.

Pd catalysts supported on CNTs and γ-Al_2O_3 were compared by Shao *et al.*[88] in the HYD of acetylene in a mixture with ethylene. The former catalyst was prepared by sonication of nitric acid-treated CNTs with a solution of Pd nitrate in ethanol, followed by evaporation of the solvent and calcining in air at 250 °C. Finally, the Pd/CNT catalyst was reduced in a flow of H_2–He (20 : 80) at 550 °C. The Pd/Al_2O_3 catalyst was available commercially. The superiority of the Pd/CNT catalyst was evident from a steady conversion and selectivity whereas over Pd/Al_2O_3 catalyst the conversion declined by about 50%. On the basis of the rate per atom of Pd, the activity of the Pd/CNT catalyst was more than five times higher than that of the Pd/Al_2O_3 catalyst.

The selective HYD of phenylacetylene to styrene under fairly mild conditions (*e.g.* 323 K, near atmospheric pressure H_2, batch process) was studied by Domínguez-Domínguez *et al.*[89] over Pd catalysts supported on CNT, CB and AC supports. The textural properties of the supports and catalysts are given in Table 7.23. These results show that compared with CNT and CB supports, the AC support favored Pd particle agglomeration. Figure 7.16[89] confirms that the best performance in terms of overall conversion was provided by the Pd/CNT catalyst. All catalysts exhibited a high selectivity with respect to styrene. After separation, the spent Pd/CNT catalyst was washed, dried and prepared for another run. Little change in the activity and selectivity of this catalyst was observed after five consecutive runs.

Table 7.23 Textural properties of supports and catalysts.[a] [89]

Sample	Pd/%	Surface area/$m^2\,g^{-1}$ Total	Surface area/$m^2\,g^{-1}$ External	Pore volume/$cm^3\,g^{-1}$ Total	Pore volume/$cm^3\,g^{-1}$ Micropores	APD/nm	Pd PS/nm
CNT		253	245	0.63	0.10	10.0	
Pd/CNT	0.66	177	177	0.48	0.07	10.8	2.5
CB		214	179	0.33	0.10	6.2	
Pd/CB	0.88	136	128	0.27	0.06	7.9	2.6
AC		2879	99	1.3	71.1	91.9	
Pd/AC	0.65	1343	52	0.66	0.56	2.0	5.6

[a]APD, average pore diameter; PS, particle size.

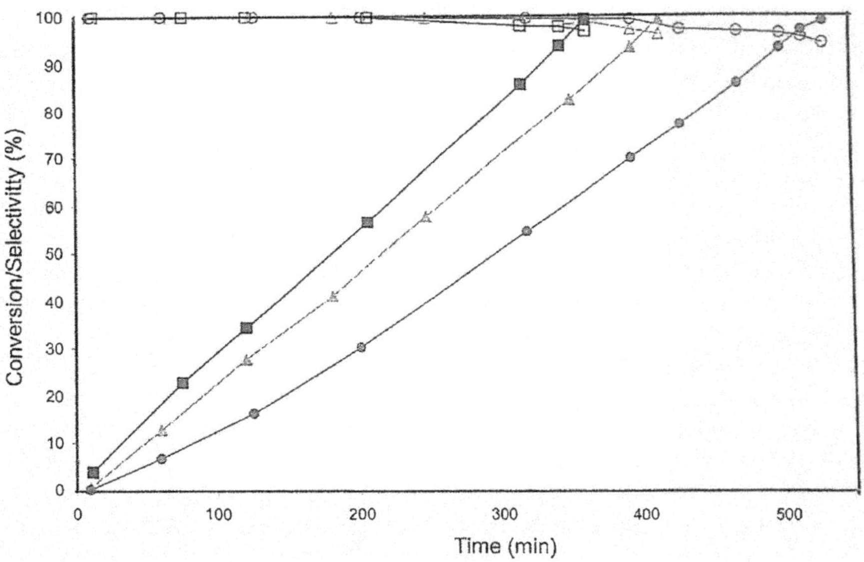

Figure 7.16 Effect of CNT, CB and AC supports on conversion and selectivity in the HYD of phenylacetylene for Pd/CNT (■, conversion; □, selectivity), Pd/CB (▲, conversion; △, selectivity) and Pd/AC (●, conversion; ○, selectivity) catalysts.
Reproduced from ref. 89 with permission from American Chemical Society, Copyright 2008.

The performance of a Pd(5%)/CNT catalyst was evaluated in the HYD of diphenylacetylene (tolane), phenylacetylene and 1-heptyne in a mixture with methanol.[90] The testing was conducted in a batch reactor at 300–333 K and 2 MPa of H_2. Complete conversion of the C≡C bond was achieved in a run lasting 2 h. The catalyst was prepared by the direct reaction of CNTs with a Pd complex such as Pd(dibenzylideneacetone)$_2$.

Liu *et al.*[91] compared Pd catalysts supported on GO, rGO and CNT with a commercial Pd/AC catalyst in the HYD of olefins (*e.g.* hexenes and cyclohexene) in ethanol and toluene under fairly mild conditions (batch system,

20 °C, 1 MPa H_2). There was a clear superiority of the Pd/GO catalyst compared with the other catalysts. The O-containing groups present on GO facilitated the efficient dispersion of Pd nanoparticles on the surface. Ethanol was a much better solvent than toluene.

The incipient wetness impregnation method was used to prepare Pd catalysts supported on CNFs and CNTs.[23] After functionalization, the supports were further modified by introducing long alkyl chains to the surface. During the HYD of butadiene (continuous fixed-bed reactor, 423 K, atmospheric pressure H_2–N_2), the Pd catalysts supported on the modified supports exhibited a high selectivity with respect to butene, which was a targeted product rather than butane. Without modification, the catalyst exhibited high activity, giving butane as the main product.

Methods for the preparation of bimetallic core–shell and alloy nanoparticles and mixtures of monometallic nanoparticles of Pt and Ru catalysts supported on CNTs were described by Li *et al.*[92] For both monometallic Pt and Ru ands bimetallic Pt–Ru catalysts, the size of the nanoparticles was in the range 1.5–3.0 nm. The activities and selectivities of the catalysts were evaluated in the HYD of phenylacetylene to styrene. The tests were carried out in a batch reactor at 50 °C and 0.4 MPa of H_2 for 1 h. The performance of the core–shell catalysts differed from that of the alloy catalysts. The conversion and selectivity of the most active Pt/CNT catalyst approached 98 and 86%, respectively, and for the next most active core–shell Pt–Ru catalyst 88 and 88%, respectively. For these catalysts, negligible catalyst deactivation ensured that the spent catalyst could be reused several times. Chen *et al.*[93] studied the HYD of 1,5-cyclooctadiene to cyclooctene at 50 °C and near atmospheric pressure H_2. For catalyst preparation, commercial CNTs were treated with nitric acid to obtain OCTNs, which were subsequently treated in a flow of $NH_3 + He$ at 400 °C to obtain NCNTs. The Pd catalysts supported on these supports were prepared by a colloidal method. After removing the solvents and drying, the solid powders obtained were reduced in a flow of H_2 (10%) + He at 200 °C. Pd nanoparticles supported on NCNTs showed a narrower size distribution than those on OCNTs. The Pd/NCNT catalyst was more active and exhibited a higher selectivity with respect to the olefin. This was attributed to the electronic effects of nitrogen on Pd, which beneficially influenced catalytic actions. At room temperature and near atmospheric pressure of H_2, a high HYD conversion of a series of unsaturated hydrocarbons (*e.g.* alkenes, cyclic di-, tri- and tetraenes, aromatic compounds, terpenes and terpenoids) was observed over Pd/CNT catalysts.[94] In addition to high selectivity, the catalysts exhibited high recyclability, as indicated by a stable catalyst activity during several subsequent tests.

Vanyorek *et al.*[95] prepared nitrogen-doped CNTs (bamboo-shaped) *via* a CVD method in a flow of N_2 using butylamine as the source of carbon and nitrogen over an Ni–MgO catalyst. Oxidizing agents such as H_2SO_4–HNO_3 mixture, concentrated HNO_3 and H_2O_2 were used for the functionalization of CNTs. The functionalized and raw CNTs were used as supports for Pd(5%) catalysts. The activities of the catalysts were compared in the HYD of octadecene to octadecane. The activity of the catalysts was determined at 50 °C and atmospheric pressure H_2 using the reactant (4.3 mL) dissolved in octanol (45.7 mL). Under these conditions, the highest HYD activity was exhibited by the catalysts supported on untreated CNTs. This was attributed to the diminished adsorption of the reactant on the highly polar surface created by the oxidation. Apparently, a much higher dispersion of Pd on the oxidized CNTs played a small role in HYD compared with the decreased adsorption of the reactant on active sites. It should be pointed out that, to a certain extent, the reactant adsorption was also inhibited by the polar solvent.

7.1.8 Summary of Studies on the HYD of Hydrocarbons

It is again emphasized that studies on the HYD of aromatics and unsaturated hydrocarbons are of significance in relation to some industrial applications. For example, HDAr of some fractions is necessary to achieve specified limits of aromatics in fuels. Ethene and styrene produced by the pyrolysis of ethane and ethylbenzene contain small amounts and acetylene and phenylacetylene, respectively, hence a polishing step is necessary to purify ethene and styrene prior to their use as feeds for the production of commercial products. Table 7.24 shows that this can be achieved over noble metal catalysts supported on CNTs under fairly mild conditions. In some studies, a relatively high H_2 pressure was used in spite of temperatures lower than 100 °C. It was suggested that a high H_2 pressure was necessary to protect active metals and thus to avoid catalyst deactivation. Studies in which catalysts supported on amorphous carbons such as ACs and CBs and also on γ-Al_2O_3 were used for comparison are of particular importance. It has been noted that in most cases, in terms of activity and selectivity, the performance of the corresponding CNT-supported catalysts was much higher than that of the catalysts supported on amorphous carbons and oxidic supports.

7.1.9 Halogenated Reactants

The conversion of halogenated compounds to hydrocarbons has attracted attention for health and safety reasons. Halogens are frequently present in waste plastics and other unwanted by-products, and must be removed to ensure environmental acceptance. Limited information on catalytic hydrodehalogenation over CNT-supported noble metal catalysts involving Br-containing reactants is available. In this regard, the novel study published by Xu *et al.*[96] on the reductive debromination of bromobenzene over Pd/CNT catalysts may be of interest in spite of the fact that hydrazine hydrate was used as a reducing agent instead of H_2. Before the addition of Pd, the CNT support was functionalized using an H_2SO_4–HNO_3 mixture. The Pd dispersion and catalyst activity were influenced by the presence of O-containing groups. The dispersion increased with increasing concentration of the O-containing groups. The experiments were conducted at 333 K by mixing either a Pd/CNT or a Pd/AC catalyst with bromobenzene and hydrazine

Table 7.24 Conversion of aromatic and unsaturated hydrocarbons.

Catalyst	Reaction	Conditions	Ref.
Aromatics			
1. Pt/CNT; Pd/CNT; Rh/CNT; Au/CNT; Pd–Rh/CNT	Benzene → cyclohexane Toluene → methylcyclohexane	Room temperature; 1 MPa; batch	79
2. Ru/CNT	Benzene → cyclohexane	308 K; 3 MPa; batch	80, 81
3. Pt/CNT; Pt/AC	Toluene → methylcyclohexane Tetralin → decalin	373 K; 0.16 MPa; continuous	82
4. Pd/CNT; Pd/AC; Pd/zeolite Y	Benzene → cyclohexane	363 K; 1 MPa; batch	83
5. Pt–Pd/CNT; Pt–Pd/ZrPSi; Pt–Pd/SiO_2–Al_2O_3	Toluene → methylcyclohexane Naphthalene → tetralin	498 K; 5 MPa; continuous	84
6. Pt/CNT; Pt/AC; Pt/Al_2O_3	Toluene → methylcyclohexane	413 K; 0.4 MPa; batch	85
7. Ru/CNT; Ru/AC; Ru/Al_2O_3	Benzene → cyclohexane	353 K; 4 MPa	86
Unsaturates			
8. Pd/CNT	Acetylene → ethylene	353–473 K; 1 atm; continuous	87
9. Pd/CNT; Pd/Al_2O_3	Acetylene → ethylene		88
10. Pd/CNT; Pd/AC; Pd/CB	Phenylacetylene → styrene	323 K; 1 atm H_2; batch	89
11. Pd/CNT	Phenylacetylene → styrene Tolane → diphenylethene Octyne → octane	300–333 K; 2 MPa; batch	90
12. Pd/CNT; Pd/GO; Pd/rGO; Pd/AC	Hexene → hexane Cyclohexene → cyclohexane	293 K; 1 MPa H_2; batch	91
13. Pd/CNT; Pd/CNF	Butadiene → butene	353–423 K; 1 atm; continuous	23
14. Pt/CNT; Ru/CNT; Pt–Ru/CNT	Phenylacetylene → styrene	323 K; 0.4 MPa; batch	92
15. Pd/CNT	Cyclooctadiene → cyclooctene	323 K; 1 atm; batch; continuous	93
16. Pd/CNT	Octadecene → octadecane	323 K; 1 atm; batch	95
17. Pd/CNT	Unsaturates → saturates[a]		94

[a]Alkenes, cyclic di-, tri- and tetraenes; aromatic compounds, terpenes and terpenoids.

hydrate in a solvent. The best performance was observed with solvents such as methanol, ethanol, 2-propanol and octane; THF and water were less efficient. Compared with Pd/AC, the Pd/CNT catalyst exhibited superior activity and selectivity.

7.2 Transition Metals Supported on CNTs

The metals that are part of conventional HPR catalysts (*e.g.* Co, Ni, Mo and W) supported on CNTs have been tested to a lesser extent than noble metals. Metal carbides and phosphides supported on CNTs have also been evaluated. Both mono- and bimetallic catalysts supported on CNTs have been investigated using reactants derived from biomass, syncrude and petroleum and also real feeds. Studies were conducted in batch and continuous systems in the presence and absence of water. Some of these metals were part of the bimetallic catalysts combined with noble metals. These catalysts were discussed in previous sections. α,β-Unsaturated aldehydes with conjugated C=C and C=O double bonds (*e.g.* CALD and citral) have received less attention compared with the extensive evaluations of these reactants over noble metal-containing catalysts. The reactants and feeds derived from different types of biomass, *i.e.* lignocellulosic, algae and vegetable oils, and also α,β-conjugated aldehydes were used to study the performance of transition metals supported on CNTs. Both partial and complete conversions to hydrocarbons have been studied.

7.2.1 Lignocellulosic Feeds and Reactants

Lignin- and cellulose- and hemicellulose-derived reactants have been included in studies using catalysts consisting of conventional metals supported on CNTs. The objectives were both partial upgrading of primary products to obtain value-added products and complete conversion to hydrocarbons.

7.2.1.1 Lignin-derived Feeds and Reactants

Lignin has attracted attention as a feed for catalytic depolymerization to obtain monomeric phenols and aromatics as reactants for further upgrading. For example, Xiao *et al.*[97] prepared an MoO_x/CNT catalyst by combining a solution of $(H_3O)_2(Mo_6Cl_8)Cl_6{\cdot}6H_2O$ with MWCNTs and subsequent calcination at 400 °C under N_2. The tests were performed in an autoclave at 260 °C and 3 MPa of H_2, using a solution of lignin in methanol (*e.g.* 50 mg lignin, 5 mg catalyst, 10 mL methanol). The yield of monomeric phenols approached 47%. In this case, a high selectivity with respect to phenolics having an unsaturated substituent was noted. This was attributed to the cleavage of the C–O bond in the β-O-4 entity (Figure 5.5) preceding HYD of the double bond as a route for the formation of such phenols. Monometallic Ni and bimetallic Ni–Me (Me = Mo, Co and Cu) all supported on nitric acid-pretreated CNTs were used for upgrading synthetic biofeed (*e.g.* a mixture of acetic acid, furfural and guaiacol) and an aqueous fraction of bio-oil in a batch reactor.[98] For the latter feed, the selectivity with respect to alcohols was of primary interest. The optimal reaction conditions were approached at 150 °C and 10 h with an initial hydrogen pressure of 5 MPa. Table 7.25[98] shows the product distribution in the upgrading of the aqueous fraction of bio-oil. Several minor products accounted for the difference from 100%. In terms of the yield of alcohols as desirable products and the H : C ratio of the liquid product, all bimetallic catalysts were more active than the monometallic Ni/CNT catalyst. Among all the catalysts tested, Ni–Mo/CNT was the most active. After the test, the Ni–Mo/CNT catalyst was recovered and reused in three subsequent runs. A loss of activity became evident during the third run.

As the product of lignin depolymerization, guaiacol has attracted attention. For example, Dongil *et al.*[99] studied the HDO of guaiacol in a batch reactor (623 K, 5 MPa, 5 h) to determine the difference in behavior between the nanosized Ni particles located inside CNTs and those on the exterior of the CNTs. For catalyst preparation, a portion of commercial CNTs was ball-milled and used as the support (CNTm) in parallel with CNTs. The Ni/CNT catalyst was prepared by stirring a CNT slurry with an ethanolic solution of Ni nitrate for 12 h and vacuum drying at 333 K. For the preparation of the Ni/CNTm catalysts, the same mixture of the support and Ni precursor was sonicated for 1 h to facilitate penetration of nanosized Ni particles into the interior of the CNTm support. The textural properties of the supports and catalysts are given in Table 7.26.[99] Transmission electron micrographs of the catalysts confirmed that a portion of the Ni particles in Ni/CNTm was located inside the CNTm. This was also confirmed by temperature-programmed reduction (TPR), which indicated that the NiO located inside can be reduced more readily. The guaiacol was completely converted over both catalysts, with the HDO approaching 25%. However, cyclohexene was the main product over the Ni/CNTm catalyst compared with cyclohexane over the Ni/CNT catalyst. In another study, Dongil *et al.*[100] prepared Ni catalysts containing variable amounts of Ni (10–20 wt%) using CNTs (as received), nitric acid-treated CNTs and ACs as supports. The activity and selectivity of

Table 7.25 Effect of catalyst type on the product distribution in the upgrading of the aqueous fraction of bio-oil.[98]

	Composition of products/%					
Catalyst[a]	Alcohols	Aldehydes	Ketones	Phenols	Acids	H : C ratio
None	7.3	15.6	23.5	15.5	28.8	0.15
Ni/CNT	42.6	8.7	9.6	8.2	23.3	0.50
Ni–Cu/CNT	55.1	5.2	5.3	6.1	21.5	0.57
Ni–Co/CNT	58.7	4.8	4.7	5.7	19.8	0.61
Ni–Mo/CNT	62.1	4.4	3.7	5.9	17.2	0.68
Ni–Mo/CNT-1	67.3	3.6	2.5	3.1	17.0	
Ni–Mo/CNT-2	62.3	4.2	3.2	4.2	17.8	
Ni–Mo/CNT-3	55.8	5.0	4.1	5.2	20.1	

[a]Conditions: 2 wt% catalyst, 150 °C, 5 MPa H_2, 10 h.

Table 7.26 Textural properties of supports and Ni catalysts.[99]

Sample	Surface area/$m^2\,g^{-1}$	Pore volume/$cm^3\,g^{-1}$		Average pore size/nm
		Micropores	Mesopores	
CNT	300	0.01	1.76	—
CNTm	375	0.02	0.44	—
Ni/CNT	241	0.01	1.08	6.1
Ni/CNTm	314	0.01	0.42	6.5

the catalysts were compared in the conversion of guaiacol at 573 K and 5 MPa of H_2 in a batch system. The catalysts were prepared by impregnation in acetone using Ni nitrate as precursor. Under the test conditions, the Ni catalyst supported on the untreated CNTs exhibited the highest initial and intrinsic rate, in line with the highest dispersion of Ni particles. The optimal amount of Ni on the catalyst approached 15 wt%. The Ni/AC catalyst was the least active. In an effort to improve the performance of the Ni(15%)/CNT catalyst further, Dongil *et al.*[101] prepared bimetallic catalysts by the addition of various amounts of Cu (1.5, 2.25, 3.0 and 3.75 wt%). The catalysts were prepared by impregnation using Ni nitrate in acetone first, followed by impregnation with Cu nitrate. The addition of Cu had an adverse effect on the intrinsic activity during the conversion of guaiacol.

The HDO of guaiacol was used as a model reaction to test the catalytic activity of monometallic Ni and Fe and bimetallic Ni–Fe catalysts in a continuous fixed-bed reactor at 573 K and 3.0 MPa H_2.[102] The catalysts were prepared by incipient wetness impregnation and co-impregnation, respectively, of commercial CNTs that had been pretreated with nitric acid. After evaporation of the solvent and drying, the catalysts were reduced in H_2 (5%) + N_2 for 4 h. As shown in Table 7.27,[102] the temperature of reduction influenced the activity and selectivity of the catalysts. In terms of hydrocarbon production and HDO rate, the best performance was exhibited by the Ni–Fe/CNT catalyst with Ni : Fe ratio = 5. The Ni/CNT catalyst was much more active than the Fe/CNT catalyst. It is believed that the activity and selectivity of Ni–Fe/CNT catalysts can be further improved by optimizing the reduction temperature and Ni : Fe ratio. With respect to selectivity, several minor products accounted for the difference from 100%.

Rahzani *et al.*[103] used an Mo/CNT catalyst for the conversion of anisole in a continuous fixed-bed system at 573–673 K and 0.8 MPa of H_2. For catalyst preparation, nitric acid-treated MWCNTs were impregnated using an aqueous solution of $(NH_4)Mo_7O_{24}\cdot 4H_2O$. After filtering and drying, the solid was calcined at 923 K under N_2, followed by reduction in a flow of N_2–H_2. Under the conditions used, benzene, phenol, 2-methylphenol, 2,6-dimethylphenol and 2,3,5,6-tetramethylphenol were formed as the main products. The product distribution suggests that the cleavage of the C_{AL}–O bond to yield phenols was favored over that of C_{AR}–O to yield benzene., Alkylation reactions giving alkylated phenols were also evident.

Table 7.27 Activity and selectivity of monometallic and bimetallic catalysts in the HYD[a] of guaiacol.[102]

Catalyst	Reduction temperature/K	Particle size/nm	Conversion/%	Selectivity/%		
				Cyclohexane	Cyclohexanol	Phenol
Ni/CNT	673	9.2	79.0	53.0	31.7	4.4
	773	13.5	80.2	24.7	58.3	1.0
	873	16.0	77.0	4.6	69.7	10.4
Fe/CNT	673	6.0	17.2	0.4	0.7	44.0
	773	6.0	19.3	0.0	1.6	92.9
	873	11.5	8.4	1.5	2.6	90.8
Ni_5–Fe_1/CNT	673	7.7	96.8	83.4	12.7	0.0
	773	8.3	96.7	67.8	23.0	0.0
	873	11.2	99.0	55.8	31.9	0.0
Ni_1–Fe_5/CNT	673	5.3	47.2	2.5	10.4	83.3
	773	5.7	41.3	2.4	13.2	79.4
	873	5.2	41.5	2.5	12.5	81.0

[a]Conditions: 573 K, 3.0 MPa H_2.

7.2.1.2 Cellulose- and Hemicellulose-derived Reactants

Yu *et al.*[104] compared monometallic Ni and Fe and bimetallic Ni–Fe catalysts supported on CNTs in the conversion of 5-HMFF in an autoclave at 473 K, 3 MPa and 3 h. The commercial CNTs were treated with nitric acid at 353 K by refluxing for 16 h. After filtering, washing and drying, the CNTs were either impregnated for monometallic or co-impregnated for bimetallic catalysts using $Ni(NO_3)_2{\cdot}6H_2O$ and $Fe(NO_3)_3{\cdot}9H_2O$ precursors. The solids produced were dried overnight at 383 K and calcined at 573 K for 4 h. For experiments, the catalysts were reduced with 5% H_2–95% N_2 at 673 K for 4 h. The bimetallic catalyst with an Ni : Fe ratio of 2 exhibited high selectivity with respect to 2,5-furandimethanol or 2,5-dimethylfuran. The spent catalyst could be readily separated with a magnet and reused several times.

Liu *et al.*[105] prepared a series of monometallic Ni and Cu catalysts and bimetallic Ni–Cu catalysts supported on CNTs using the impregnation method. The catalysts were used in the HYD of furfural (FF) to tetrahydrofurfuryl alcohol (THFFA) in a batch reactor at 100 and 130 °C under H_2 pressures varying from 1 to 4 MPa. Catalysts with similar metal loadings but supported on oxidic supports such as Al_2O_3, TiO_2, ZrO_2 and MgO were also prepared and tested for comparison. In all respects, the CNT-supported catalysts outperformed those supported on oxidic supports. Table 7.28[105] gives the results obtained over catalysts supported on CNTs at 100 °C with a 10 h duration. An increase in H_2 pressure from 1.0 to 3.0 MPa resulted in an increase in conversion and a significant decrease in the formation of the unwanted 2-furaldehyde diethyl acetal (FDA) product. At the same time, THFFA became the predominant product. A temperature increase from 100 to 130 °C had little effect on catalyst performance.

Table 7.28 Effect of H_2 pressure on conversion and selectivity in the HYD of furfural over different catalysts.[105]

			Selectivity[a]/%			
Catalyst	Pressure/MPa	Conversion/%	FFA	THFFA	FDA	Others
Ni(5%)/CNT	1	62.7	24.3	21.5	49.4	4.8
	3	94.9	52.3	38.5	3.5	5.7
	4	96.7	48.9	40.4	5.2	5.5
Ni(10%)/CNT	1	70.1	14.3	36.7	38.2	10.8
	3	98.4	3.9	83.4	5.6	7.1
	4	98.7	2.6	84.9	7.1	5.4
Ni(10)%/CNT[b]	4	99.1	2.1	85.1	6.8	6.0
Cu–Ni/CNT	1	68.7	8.9	46.3	32.7	12.1
	3	98.8	1.9	85.3	1.1	11.7
	4	99.8	1.0	88.1	2.1	8.8
Cu–Ni/CNT[a]	4	100	0.2	90.3	1.4	8.1

[a]FFA, furfuryl alcohol; THFFA, tetrahydrofurfuryl alcohol; FDA, 2-furaldehyde diethyl acetal.
[b]Reaction temperature 130 °C.

A nonsulfided Ni catalyst supported on nitric acid-functionalized CNTs was used for the HDO of bio-model compounds such as acetic acid, furfural and hydroxyacetone and also raw bio-crude.[106a] The catalytic tests were performed in an autoclave in the temperature range 100–250 °C. The addition of Mo enhanced the activity, and the bimetallic Ni(10%)–Mo(10%)/CNT catalyst was the most active. Similar results were also obtained in the upgrading of bio-crude, *e.g.* the hydrogen content of the product increased from 8.4 to 11.2%.

Gong *et al.*[106b] prepared N-doped CNTs *via* pyrolysis of melamine as a source of both carbon and nitrogen. The CNTs were used for the preparation of Co/CNT and Ni/CNT catalysts by a method ensuring the embedding of the active metals. The catalysts exhibited remarkable activity and selectivity in the HYD of furfural to furfuryl alcohol in the aqueous phase. For example, almost complete conversion of furfural to tetrahydrofurfuryl alcohol was achieved at 100 °C and 4 MPa of H_2. The catalysts retained high activity and selectivity during several subsequent cycles.

Mai *et al.*[107] compared CNTs and ACs as supports for Mo_2C catalysts in the HYD of levulinic acid (LA) to γ-valerolactone (GVL), as shown in Figure 5.4. The tests were conducted in a continuous reactor at 100–200 °C and 2 MPa. The advantage of the CNT support at all temperatures was evident. Spectroscopic evaluation revealed that Mo_2C crystallites were located in the interior of the CNTs. This ensured their stability, as indicated by the absence of deactivation. Quiroz *et al.*[108] prepared well-defined β-Mo_2C from MoO_3 deposited on MWCNTs by heating at 700 °C in an inert atmosphere. In this case, the MWCNTs served as the source of carbon. Another β-Mo_2C was prepared by carburization of the MoO_3 deposited on MWCNTs at 650 °C using a CH_4–H_2 mixture. The activity and selectivity of the catalysts in the

HYD of LA to GVL in the aqueous phase (batch system, 180 °C, 3 MPa of H_2) were determined. The catalyst prepared by carburization was more active, with selectivity with respect to GVL exceeding 85%, and 1,4-pentanediol and methyltetrahydrofuran were minor products.

7.2.2 Vegetable Oils and Algae-derived Reactants

Liang *et al.*[109] reported that a β-Mo_2C/CNT catalyst exhibited superior activity and selectivity in the conversion of model compounds typical of those present in microalgae. Thus, stearic acid and hexadecanamide were almost completely converted to *n*-C_{15}–C_{18} alkanes with more than 90% selectivity with respect to *n*-octadecane. The β-Mo_2C/CNT catalyst was prepared by the carbothermal hydrogen reduction method. The tests were conducted in an autoclave at 180 °C, 4 MPa of H_2 and 4 h. The catalyst maintained a high stability, as confirmed by only a slight decrease in activity in six subsequent cycles.

Han *et al.*[110] used MWCNTs for the preparation of a Mo_2C(20%)/CNT catalyst to be used in the HDO and isomerization of vegetable oils. In the first step, CNTs were treated with nitric acid at 130 °C and the filtered solids were dispersed in an aqueous solution of ammonium molybdate. After sonication and evaporation of water, the solid was dried at 120 °C and carburized at 600 °C in a flow of H_2. Pd(5%)/CNT and Pt(5%)/CNT catalysts were prepared by the incipient wetness impregnation method using the same CNTs. In this case, solutions of H_2PdCl_4 and H_2PtCl_4, respectively, were used as precursors. The performances of these catalysts under the same conditions are compared in Table 7.29.[110] It is evident that Mo_2C(20%)/CNT was much more active than Pd(5%)/CNT and Pt(5%)/CNT catalysts. The Mo_2C/CNT catalyst exhibited a high selectivity for C_{16} hydrocarbons whereas the Pd/CNT and Pt/CNT catalysts were selective for decarbonylation, giving pentadecane as the main product. Sunflower oil in hexane was used as the feed to evaluate hydroisomerization activity under the same conditions as above. Among the catalysts tested, the nanosized Mo_2C(20%)/CNT catalyst exhibited the best performance. As the results in Figure 7.17[110] show, over this catalyst a high content of isoalkanes was present in the products from the conversion of several vegetable oil feeds. Moreover, the catalyst was

Table 7.29 Effect of catalyst composition on product distribution.[110]

		Product distribution[b]/%				
Catalyst[a]	Conversion/%	A	B	C	D	Selectivity[c]/%
Mo_2C(10%)/CNT	72	2.6	49	26	0.8	78
Mo_2C(20%)/CNT	90	2.1	57	32	0.3	91
Mo_2C(40%)/CNT	86	2.1	50	36	0.5	88
Pd(5%)/CNT	41	84	11	2.1	0.2	87
Pt(5%)/CNT	32	70	11	3.0	0.4	84

[a]Conditions: 240 °C, 1.5 MPa, 2 h.
[b]A, pentadecane, B, hexadecane, C, hexadecene, D, hexadecanal.
[c]Selectivity is the sum of the yields of hydrocarbons.

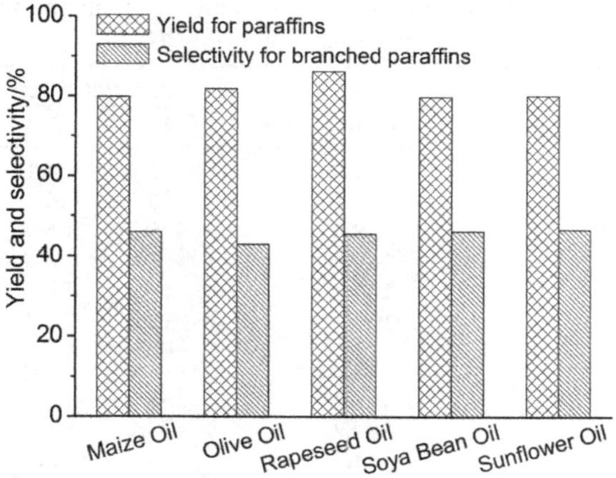

Figure 7.17 Effect of feed origin on the yield of *n*-alkanes and isoalkanes over Mo_2C/CNT catalyst.
Reproduced from ref. 110 with permission from the Royal Society of Chemistry.

much more resistant to the loss of active metals due to leaching compared with the Pd/CNT and Pt/CNT catalysts.

Ding *et al.*[111] compared catalysts such as MoO_2(5%)/CNT with Pd(5%)/CNT in the HDO of palmitic acid. The former catalyst was prepared by mixing a solution of $(NH_4)_6Mo_7O_{24}\cdot 4H_2O$ in water with CNTs. The paste was dried at 105 °C, tempered at 500 °C and reduced in H_2 at 500 °C for 4 h. The Pd/CNT catalyst was prepared by incipient wetness impregnation. Catalyst evaluation was conducted in an autoclave at 190–260 °C and 4 MPa of H_2. Typically, 0.5 g of reactant in decane and 0.25 g of catalyst were used. The results are compared in Table 7.30.[111] The predominance of hexadecane over MoO_2/CNTs compared with pentadecane over Pd/CNTs indicated two different HDO pathways. Thus, the former catalyst gave a better carbon yield by minimizing decarboxylation. On the other hand, the loss of one carbon due to decarboxylation, giving pentadecane, was the dominant reaction over Pd/CNT. At the end of test, the spent MoO_2(5%)/CNT catalyst was recovered by filtration, washed, dried and reused in further runs. After five repeated runs, little catalyst deactivation was observed. In another study conducted by Ding *et al.*[112] under identical conditions, the MoO_2/CNT catalyst was promoted by Co. This resulted in a significant increase in catalyst activity, whereas little effect on the reaction pathway was observed. Thus, hexadecane was predominant product.

Chen *et al.*[113] prepared Co/CNT, Ni/CNT and Pt/CNT catalysts, each containing about 5 wt% of metal, by incipient wetness impregnation of CNTs using aqueous solutions of the metal nitrates. After drying, the samples were calcined at 500 °C under N_2, followed by reduction in H_2 at 700 °C. The catalysts were used for the hydroliquefaction of algae biomass such as *Dunaliella tertiolecta* to produce bio-crude. Tests were conducted in an

Table 7.30 Effect of temperature on conversion and selectivity in the hydrogenation of palmitic acid over MoO_2/CNT and Pd/CNT catalysts.[111]

			Product distribution[b]/%			
Catalyst[a]	Temperature/°C	Conversion/%	A	B	C	D
MoO_2/CNT	190	53.8	2.4	4.1	47.3	0
	200	96.1	5.8	20.1	70.0	0
	210	98.6	5.9	64.9	27.7	0.2
	220	100	7.6	92.2	0	0.2
	240	100	14.2	78.7	0	7.1
	260	100	15.4	72.4	0	12.2
None	220	10.3	3.4	2.0	0	4.9
Pd/CNT	220	64.2	58.0	2.0	4.2	0
	240	80.2	74.1	2.5	2.1	1.5
	260	93.3	85.4	3.1	0	4.8
	280	95.8	78.3	12.1	0	5.4

[a]Conditions: 0.5 g of palmitic acid in 50 mL of decane, 0.25 g of catalyst, 4 MPa H_2, 4 h.
[b]A, pentadecane; B, hexadecane; C, hexadecanol; D, light alkanes.

autoclave at 320 °C and 12 MPa of H_2. With respect to the quality of bio-crude, the best performance was exhibited by the Co/CNT catalyst. Thus, hydrocarbons were the predominant products, with fatty acids and N-containing compounds accounting for a smaller portion of the bio-crude, in contrast to the products obtained over Ni/CNT and Pt/CNT catalysts.

Duan *et al.*[114] prepared a Ni(5%)/CNT catalyst by incipient wetness impregnation of CNTs at room temperature. After separation, the paste was dried at 105 °C for 12 h and reduced in H_2 at 500 °C for 4 h. The catalyst was used for the conversion of palmitic acid to hydrocarbons in a batch reactor at 240 °C and 2 MPa of H_2. Under these conditions, the conversion *via* decarbonylation approached 97% with 90% selectivity with respect to pentadecane.

7.2.3 Conversion of α,β-Unsaturated Aldehydes

Malobela *et al.*[115] compared MWCNTs with graphite and ACs as supports for Co–Ni catalysts in the conversion of CALD to CALC (Figure 5.2). The experiments, lasting 10 h, were conducted in a mini-autoclave at 150 °C and 0.5 MPa. The Co–Ni/MWCNT catalyst exhibited the best performance. For this catalyst, high activity facilitated by Ni was combined with high selectivity with respect to alcohol after the addition of Co. For example, at a conversion of 63%, the product contained 62% of CALC and 38% of HCALD.

Bimetallic catalysts comprising either Cu combined with Fe, Co, Ni, Zn or Cu–Ru supported on MWCNTs and a mesoporous carbon (MC) were evaluated in the HYD of CALD by Hareesh *et al.*[116] The catalysts supported on the CNTs exhibited better performance than those supported on MC. In terms of conversion and selectivity with respect to CALC, the best catalyst was Cu(10%)–Ru(1%)/CNT. The conversion and selectivity approached 85% and almost 80%, respectively.

Tang *et al.*[117] compared an Ni catalyst supported on poly(acrylic acid)-grafted CNTs with the Ni/AC counterpart in the HYD of citral (Figure 5.3). The experiments were conducted in a batch reactor at 80 °C and 1 MPa of H_2. For catalyst preparation, Ni particles were deposited on the CNTs by chemical reduction using hydrazine. The objective was to obtain a high selectivity for cintronellal, used in the flavoring and perfumery industries. At 73% citral conversion, the selectivity with respect to citronellal exceeded 90% and geraniol and citronellol accounted for the rest. Under the same conditions, at 90% overall conversion over Ni/AC, the selectivity was only 11%.

7.2.4 Other Oxygenates

Perhaps the first information on the preparation and testing of cobalt silicides (CoSi) supported on CNTs was published by Zhang *et al.*[118] For catalyst preparation, CNTs, $CoCl_2$ and silicon powder were mixed in an agate mortar and the homogenized mixture was transferred into a reactor and fluidized with argon to expel oxygen. Subsequently, the solid was microwaved to obtain CoSi/CNT catalyst. The performance of this catalyst was evaluated in the HYD of phthalic anhydride at 180–220 °C and 4 MPa in a batch reactor. For testing, the phthalic anhydride was mixed with hydroxybutanoic acid lactone.

7.2.5 Summary of Studies on the Conversion of Oxygenated Reactants and Feeds Over Transition Metals on CNTs

In general agreement with noble metal catalysts supported on CNTs (Tables 7.8 and 7.22), the conversion of oxygenates to hydrocarbons over transition metal catalysts supported on CNTs required more severe conditions than partial conversions, as shown in Table 7.31. Such conditions approach those encountered during HPR. The depolymerization of lignin to monomers requires more severe conditions because several C–O bonds must be cleaved before a monomer can be released. Although similar results for cellulose and hemicellulose are not available, milder conditions may be anticipated than those employed in the depolymerization of lignin. Thus, it is well known that lignin is a much more stable component of lignocellulosic biomass than cellulose and hemicellulose.

Much less attention has been paid to the HYD of α,β-unsaturated aldehydes (*e.g.* CALD and citral) and other oxygenates over conventional metal-containing catalysts compared with the extensive efforts involving noble metals catalysts. Thus, based on Tables 7.8, 7.22 and 7.31, the total number of studies involving the former catalysts approached 20 compared with 72 studies that were conducted over noble metal catalysts supported on CNTs. Apparently, high catalyst selectivity combined with high conversions could be achieved more readily over the latter catalysts. Most of the studies were carried out in batch systems. For noble metal catalysts supported on CNTs, a unique selectivity, stemming from mutual effects between the support and active metals, resulted in the formation of fine chemicals and

Table 7.31 Conversion of oxygen-containing reactants and feeds over transition metal catalysts supported on CNTs.

Catalyst	Feed/reactant	Conditions	Ref.
Partial conversion			
1. Mo_2C/CNT	Lignin → monomers	533 K; 3 MPa; methanol	97
2. Ce/CNT; La/CNT	Lignin → monomers	533 K; 4 MPa; batch	139
3. Ni/CNT; Ni–Me/CNT (Me = Mo, Co, Cu)	Bio-oil → alcohols	423 K; 5 MPa; aq.	98
4. Ni/CNT; Fe/CNT; Ni–Fe/CNT	5-HMF → 2,5-DMF	473 K; 3 MPa	103
5. Ni/CNT; Cu/CNT; Ni–Cu/CNT; Ni, Cu supported on Al_2O_3, TiO_2, ZrO_2 and MgO	FF → THFFA	373–403 K; 1–4 MPa	104
6. Ni/CNT; Ni–Mo/CNT	FF, bio-oil → alcohol	373–523 K; 1–4 MPa	105
7. Mo_2C/CNT; Mo_2C/AC	LA → GVL	373–473 K; 2 MPa; aq.	107
8. Mo_2C/CNT	LA → GVL	453 K; 3 MPa; aq.	107
9. Co–Ni/CNT, Co–Ni/AC; Co–Ni/graphite	CALD → CALC	423 K; 0.5 MPa; batch	115
10. Cu–Ru/CNT; Cu–Me/CNT (Me = Fe, Co, Ni, Zn)	CALD → CALC	373 K; 1 atm H_2; batch; ethanol	116
11. Ni/CNT; Ni/AC	Citral → alcohol	353 K; 1 MPa; batch; ethanol	117
Conversion to hydrocarbons			
12. Ni/CNT	Guaiacol	623 K; 5 MPa; batch	99
13. Ni/CNT; Ni–Cu/CNT; Ni/AC	Guaiacol	573 K; 3 MPa; continuous	100, 101
14. Ni/CNT; Fe/CNT; Ni–Fe/CNT	Guaiacol	573 K; 3 MPa; continuous	102
15. Mo/CNT	Anisole	573–673 K; 0.8 MPa	108
16. Mo_2C/CNT	Stearic acid	453 K; 4 MPa; batch	109
17. MoO_2/CNT; Pd/CNT; Pt/CNT	Vegetable oil	513 K; 1.5 MPa; batch; hexane	110
18. MoO_2/CNT; Pd/CNT; $CoMoO_2$/CNT	Palmitic acid	463–553 K; 4 MPa; batch; decane	111, 112
19. Co/CNT; Ni/CNT; Pt/CNT	Algae bio-oil	593 K; 12 MPa; batch	113
20. Ni/CNT	Palmitic acid	513 K; 2 MPa; batch	114

intermediates of industrial importance. Therefore, over these catalysts, complete conversion to fuel hydrocarbons has attracted less attention.

7.2.6 Petroleum-derived Reactants and Feeds

Studies of aromatics and heteroring-containing reactants typical of those present in conventional petroleum in addition to studies involving real feeds derived from petroleum have appeared in the literature.

7.2.6.1 Conversion of Aromatics

Sun *et al.*[119] prepared Ni/CNT catalysts by impregnation in a water–ethylene glycol solution of Ni precursor. Another Ni/CNT catalyst was prepared by the same method but in the absence of ethylene glycol. These catalysts are denoted Ni/CNT-e and Ni/CNT-n, respectively. The latter catalyst was reduced in H_2 at 400 °C to obtain Ni/CNT-h catalyst. In parallel, Ni/AC-e and Ni/AC-n catalysts were prepared by the same method. These catalysts were evaluated in the HYD of benzene in a continuous reactor at near atmospheric pressure H_2 at 160 °C. The activities of the catalysts are compared in Figure 7.18.[119] It is evident that the presence of ethylene glycol was necessary to obtain the highest activity Ni/CNT-e catalyst. The corresponding Ni/AC-e catalyst was much less active.

Yang *et al.*[120] prepared two types of Ni/CNT catalysts, *i.e.* one with Ni filled inside the channels of the CNTs and the other on the external surface of the CNTs. For the former, MWCNTs (od 8–20 nm and id 4–8 nm) were ball-milled to shorten the length and open the ends before being purified using nitric acid. After washing and drying, the CNTs were mixed with a solution of

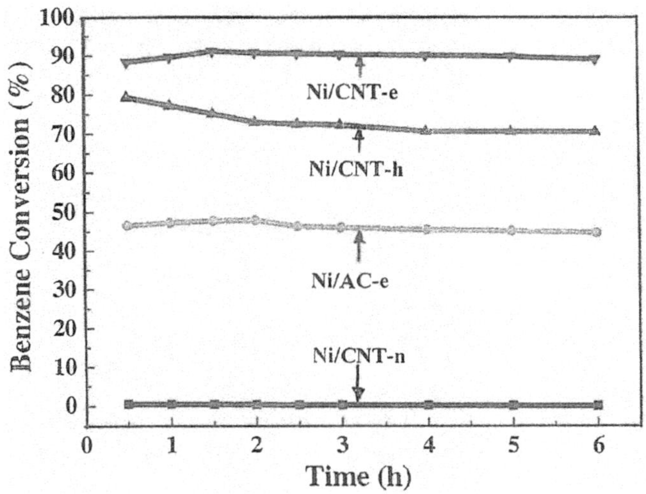

Figure 7.18 Effect of catalyst preparation conditions on the activity of Ni/CNT catalysts in benzene conversion.
Reproduced from ref. 119 with permission from Elsevier, Copyright 2016.

$Ni(CH_3COO)_2{\cdot}6H_2O$ in acetic acid. The Ni-containing liquid entered the channels of the CNTs with the aid of capillary force. After drying, the solid was treated at 450 °C under N_2. The final form of the catalyst was obtained by reduction in H_2 at 500 °C for 3 h. For the preparation of Ni/CNT catalysts with Ni particles deposited on the external surface of CNT, the unopened CNTs were treated with nitric acid before being immersed in Ni acetate solution. The paste obtained was treated as above to obtain the final form of Ni/CNT catalysts. The catalysts were applied in the HYD of benzene to cyclohexane. The testing of these catalysts was conducted in a continuous microreactor at 150 °C and atmospheric pressure H_2. Figures 7.19 and 7.20[120] confirm that for both reduced and unreduced catalysts, respectively, a higher activity was

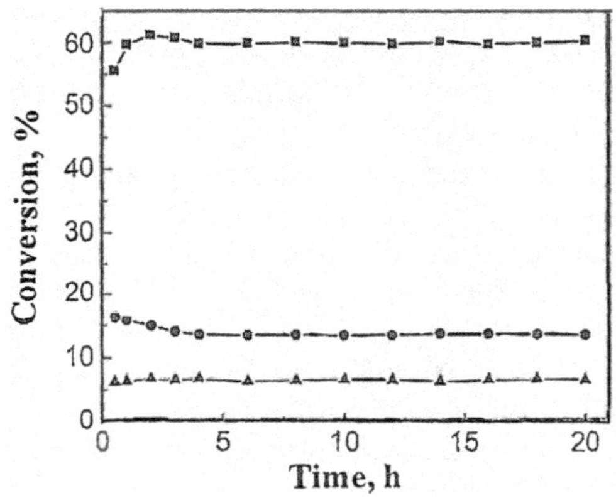

Figure 7.19 Effect of Ni location on the activity of Ni/CNT catalysts (reduced) during benzene conversion: ■, Ni in CNTs; ●, Ni on CNTs; ▲, Ni on AC. Reproduced from ref. 120 with permission from Elsevier, Copyright 2010.

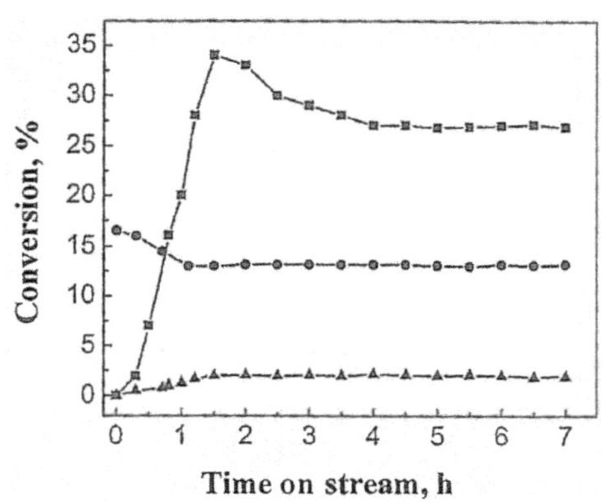

Figure 7.20 Effect of Ni location on the activity of Ni/CNT catalysts (unreduced) during benzene conversion: ■, Ni in CNTs; ●, Ni on CNTs; ▲, Ni on AC. Reproduced from ref. 120 with permission from Elsevier, Copyright 2010.

exhibited by the catalyst with Ni particles in the channels than that with Ni on the exterior of the CNTs. This was attributed to a higher density of H_2 and benzene inside the tubes than that on the outside of the tubes. The beneficial effect of reduction was the most evident for the for the Ni/CNT catalyst with the Ni particles in the channels. The Ni/AC catalyst was much less active.

Ju and Li[121] compared an NiP amorphous alloy alone with NiP supported on CNTs in the HYD of benzene to cyclohexane. The catalysts contained 91 and 16.5 wt% of Ni, respectively. Testing was conducted in a continuous system between 393 and 433 K and near atmospheric pressure H_2. Figure 7.21[121] shows a higher overall conversion over the NiP alloy than over Ni/CNT. However, Ni/CNT was much more active when the conversion was expressed per unit of Ni. Moreover, the stability of the NiP/CNT catalyst was also much higher.

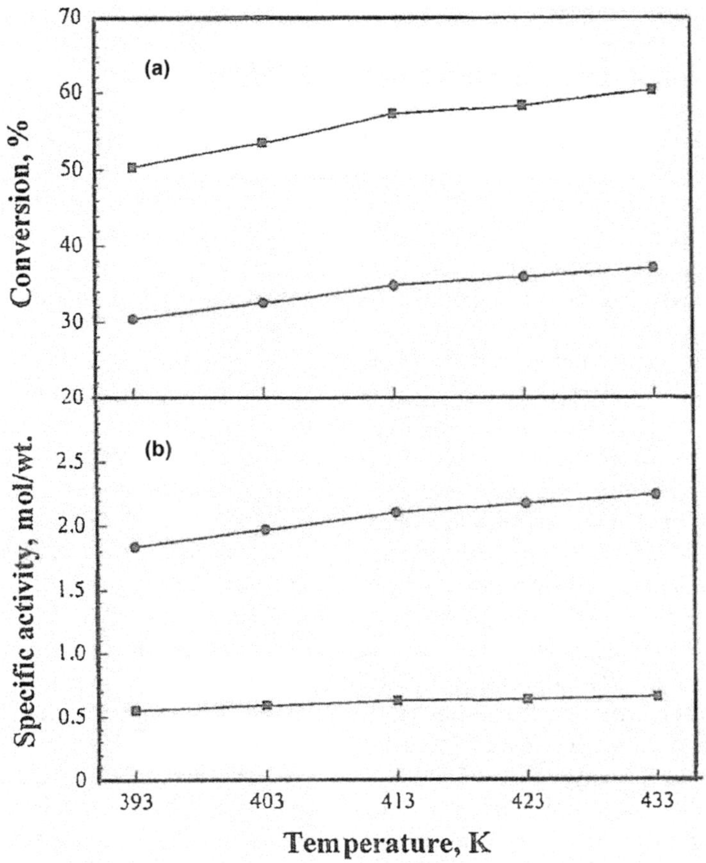

Figure 7.21 (a) Benzene conversion and (b) specific activities over (■) NiP amorphous alloy and (●) Ni/CNT catalyst.
Reproduced from ref. 121 with permission from Elsevier, Copyright 2006.

7.2.6.2 Conversion of Heteroaromatics

The promoting effect of Ni on the HDS activity of MoS_2 observed in conventional HPR catalysts[76,77] was also confirmed for Ni–MoS_2 catalysts supported on CNTs in a study by Liu *et al.*[122] They used the HDS of thiophene as a model reaction. The catalyst preparation involved aqueous-phase impregnation of nitric acid-functionalized CNTs using $(NH_4)_6Mo_7O_{24}{\cdot}24H_2O$ and $Ni(NO_3)_2{\cdot}6H_2O$ precursors. In addition, single Ni and Mo metal catalysts were prepared for comparison. After drying and calcining, the catalysts were sulfided with CS_2 dissolved in hexane. The HDS activity was determined in a continuous fixed-bed reactor at 180–240 °C in a flow of H_2 using thiophene dissolved in *n*-heptane. The following order of HDS activity was established: Ni–Mo/CNT ≫ Mo/CNT > Ni/CNT. The HDS activity of a mechanical mixture of Mo/CNT and Ni/CNT exceeded that of the monometallic catalysts but was lower than that of the Ni–Mo/CNT catalyst.

Nejad *et al.*[123] used a solution of thiophene (1 wt%) in decane as the feed to evaluate the performance of trimetallic Ni (4%), Mo (12%) and W (0, 3, 6 and 9%) catalysts supported on MWCNTs. For the catalyst preparation, a solution containing Mo, W and Ni was added to a slurry of CNTs together with ammonia solution followed by ultrasonic treatment at 70–80 °C. After filtration, washing and drying, the solid was treated under Ar at 450 °C, followed by sulfidation with an H_2–H_2S mixture. The experiments were carried out in a batch reactor at 160 °C and near atmospheric pressure H_2. The best performance was exhibited by the catalyst containing 6 wt% of W. This catalyst had a small particle size with a uniform dispersion on the support.

Dong *et al.*[124] compared the activities of Co–Mo/CNT and Co–Mo/AC catalysts in the HDS of thiophene and HDN of pyrrole in a continuous system at 613 K and 1.5 MPa of H_2. During catalyst preparation, nitric acid-treated CNTs were impregnated using an aqueous solution of $Co(NO_3)_2{\cdot}6H_2O$ first, followed by impregnation using ammonium molybdate solution. After calcining at 773 K under N_2, the catalysts were sulfided in CS_2–H_2 (5 : 95) mixture. A series of catalysts containing ~6 wt% Mo but different amounts of Co, *i.e.* giving Co : Mo ratios varying from 0 to 3, were prepared. In HDS, the best performance was observed at an Mo : Co ratio of 3. At this ratio, the activity was further increased by increasing the amount of Mo from 6 to 7.2 wt%. The activity difference in Figures 7.22 and 7.23[124] confirmed the superior performance of the Co–Mo/CNT catalysts compared with Co–Mo/AC catalysts in both the HDS of thiophene and HDN of pyrrole.

The HDS activity of Co–Mo/CNT catalysts was evaluated using a model gasoline feed containing a mixture of cyclohexane, diisobutene, cyclohexene and 1-octene in a volume ratio of 60 : 30 : 5 : 5.[125] The mixture was spiked with 0.5% of thiophene. At 260 °C and near atmospheric pressure H_2, complete HDS of thiophene was observed. The Co : Mo atomic ratio had a pronounced effect on the catalyst performance and the optimal conditions were attained at a ratio of 0.4.

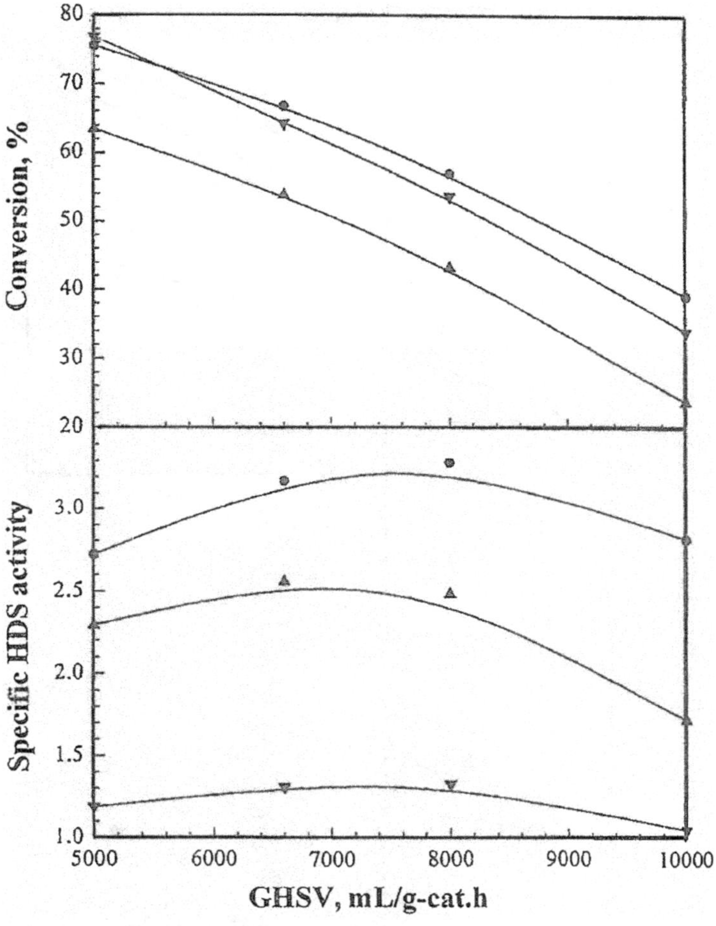

Figure 7.22 Effect of GHSV on the conversion of thiophene (613 K; 1.5 MPa): ●, $7.24Mo_3Co_1/CNT$; ▲, $7.24Mo_3Co_1/AC$; ▼, $16.9Mo_3Co_1/AC$. Reproduced from ref. 124 with permission from Elsevier, Copyright 2006.

Shang and co-workers[126–128] studied the activity of sulfided Co–Mo catalysts supported on CNTs in comparison with the corresponding catalysts supported on γ-Al_2O_3 in the HDS of dibenzothiophene (DBT) in a continuous fixed-bed reactor at 280 °C, 2 MPa and an LHSV of 6 h^{-1}. The catalysts were prepared by pore volume impregnation with the Mo precursor first followed by that of the Co precursor. After drying at 110 °C, the catalysts were heat treated under N_2 at 500 °C. The catalysts were sulfided in a flow of H_2–H_2S (5%) at 400 °C. For HDS, a 1% molar solution of DBT in toluene was used. Table 7.32[127] indicates a higher activity of the Co–Mo/CNT catalysts compared with the Co–Mo/Al_2O_3 catalyst. The TPR data confirmed than an oxidic form of Co–Mo/CNT could be reduced much more readily than that of

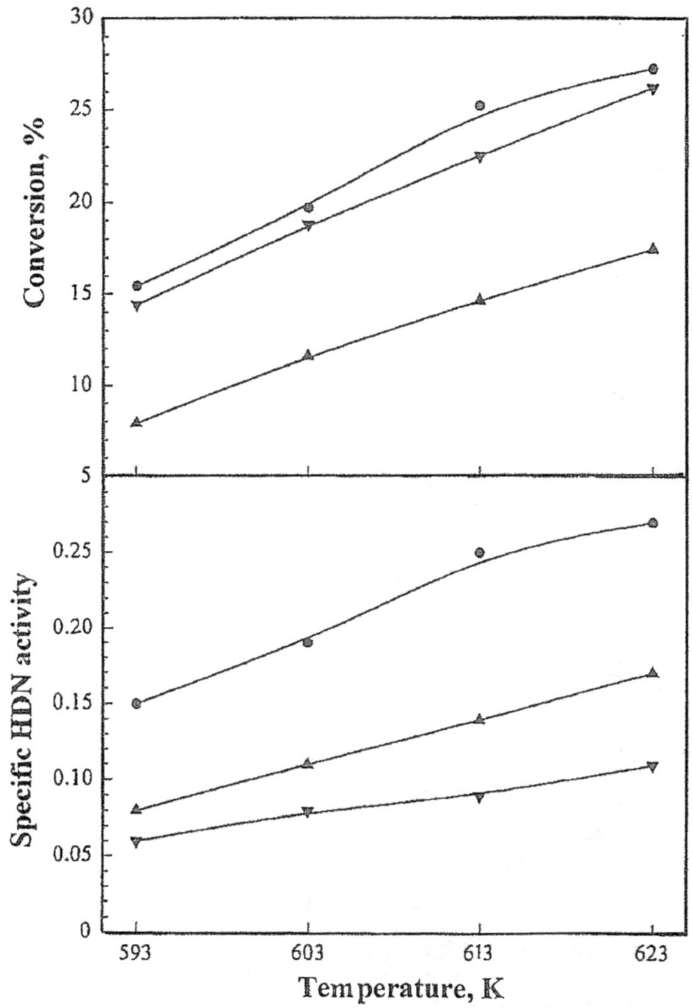

Figure 7.23 Effect of temperature on the HDN of pyrrole: ●, 7.24Mo_3Co_1/CNT; ▲, 7.24Mo_3Co_1/AC; ▼, 16.9Mo_3Co_1/AC.
Reproduced from ref. 124 with permission from Elsevier, Copyright 2006.

Co–Mo/Al_2O_3. Sulfiding of the catalysts was performed at 340 °C and 2 MPa of H_2 for 4 h, using a mixture of 3 mol% CS_2 in cyclohexane.

Whelan *et al.*[129] prepared a series of Co–Mo/CNT catalysts using an incipient wetness impregnation method. The first catalyst was prepared by sequential impregnation of the CNT support by Mo followed by Co. The next catalyst was prepared by impregnation with Co first followed by Mo. In addition, co-impregnation with Co and Mo was also performed. After drying, the solids were calcined at 350 °C. After sulfiding, the catalysts were tested in the HDS of DBT (1% DBT in decalin) in a batch reactor at 300 °C and 3.5 MPa

Table 7.32 Product distributions in the HDS reaction of DBT over Co–Mo(x) catalysts, where $x = $ Co : Mo ratio.[127]

Catalyst[a]	Product[b]/%			HDS/%
	CHB	BPN	4H-DBT	
Co–Mo(0.35)/γ-Al_2O_3	15.0	68.0	0.6	83.0
Co–Mo(0.2)/CNT	7.2	87.0	0.5	94.2
Co–Mo(0.35)/CNT	5.5	91.2	0.5	96.7
Co–Mo(0.5)/CNT	7.2	88.8	0.1	96.0
Co–Mo(0.7)/CNT	8.8	90.8	0.4	99.6
Co–Mo(1.0)/CNT	7.5	83.7	0.6	91.2

[a]Conditions: 280 °C, 2 MPa, LHSV = 6 h^{-1}.
[b]CHB, cyclohexylbenzene; BPN, biphenyl; 4H-DBT, tetrahydrodibenzothiophene.

of H_2. The highest HDS activity was exhibited by the catalyst prepared by sequential impregnation with Mo first, followed by the catalyst prepared by co-impregnation then that prepared that by impregnation with Co first, *i.e.* 82, 64 and 60%, respectively.

7.2.7 Real Petroleum Feeds

CNTs prepared and functionalized by treatment with HNO_3–H_2SO_4 solution were used by Kiani *et al.*[130] as supports for the preparation of Ni_2P/CNT catalysts. For the preparation of the catalysts, CNTs were mixed with an aqueous solution of Ni dihydrogenphosphite. The deposition of nickel phosphide (Ni_2P) on functionalized CNTs decreased the surface area compared with little change for the nonfunctionalized CNTs. This was attributed to the filling of the pore cavities of the CNTs. The solid obtained was dried, pelletized and heated at 375 °C in a flow of H_2. Catalysts containing 5, 10 and 15 wt% Ni_2P on CNTs were prepared and tested in the HDS of naphtha (3000 ppm of S) in a continuous fixed-bed reactor. The S content in the products decreased to 2100, 280 and 520 ppm, respectively. Therefore, Ni_2P(10%)/CNT was the most active. For this catalyst, a decrease in S content to 41 ppm was achieved by optimizing the conditions. Ni_2P supported on functionalized CNTs exhibited much higher activity.

Soghrati *et al.*[131,132] used Co-Mo-impregnated cordierite ($2Al_2O_3$–$5SiO_2$–$2MgO$) monolith as the catalyst for the preparation of CNTs by decomposition of methane in a mixture with H_2 at 900 °C. The monolith coated with CNTs was oxidized in a stream of ozone followed by oxidation using 30% H_2O_2 solution. This step removed Co and Mo metals from the monolith. The pretreated monolith and CNT-covered monolith were impregnated using aqueous solutions of $Co(NO_3)_3$ and $(NH_4)_6Mo_7O_{24}\cdot 24H_2O$ to obtain the corresponding Co–Mo catalysts. After impregnation, the catalysts were dried at 120 °C and calcined at 600 °C under N_2. The H_2 prereduced catalysts were sulfided using 1% dimethyl disulfide (DMDS) in a hydrocarbon liquid at 310 °C under an H_2 pressure of 3 MPa. The Co–Mo/monolith and Co–Mo/CNT-covered monolith catalysts, together with a conventional Co–Mo/Al_2O_3

catalyst, were used in the HDS of a naphtha fraction derived from an Iranian crude. The testing was conducted in a continuous fixed-bed reactor at 310 °C, 1.5 MPa of H_2 and an LHSV of 4 h^{-1}. Among the three catalysts tested, the Co–Mo/CNT-covered monolith exhibited the highest HDS activity.

Carbon materials such as carbon blacks (CBs), graphene nanoplatelates (GNPs) and CNTs were compared with a zeolite as supports for Ni catalysts in a study conducted by Guo *et al.*[133] The size of the Ni particles in all the catalysts was ~12 nm. The catalysts were used in the upgrading of a heavy feed. Experiments of 2 h duration were conducted in a batch reactor at 300 °C and 2–4 MPa. With respect to the decrease in the viscosity of the heavy feed, the following order of catalyst activity was established: no catalyst < Ni/zeolite < Ni/GNP < Ni/CNT < Ni/CB. Under similar conditions, the unpromoted CB was also slightly more active than the other supports.

Co–Mo/CNT and Co–Mo/Al_2O_3 catalysts were compared in the HDS of naphtha feed containing 1200 ppm of S.[134] The catalysts were prepared by incipient wetness impregnation using a solution of $(NH_4)_6Mo_7O_{24}{\cdot}24H_2O$ first, followed by the $Co(NO_3)_2$ precursor. Subsequently, the catalysts were treated at 773 K in N_2 for the Co–Mo/CNT and in air for the Co–Mo/Al_2O_3 catalyst. The Co : Mo atomic ratio of these catalysts was 0.35. The catalysts were first reduced in H_2 and then sulfided using a solution of dimethyl sulfide in hexane. The catalysts were evaluated in a continuous system at 533–623 K and 1.5 MPa of H_2 using a naphtha feed containing 1200 ppm of S. Figure 7.24[134] shows a better performance of the Co–Mo/CNT catalyst over the entire

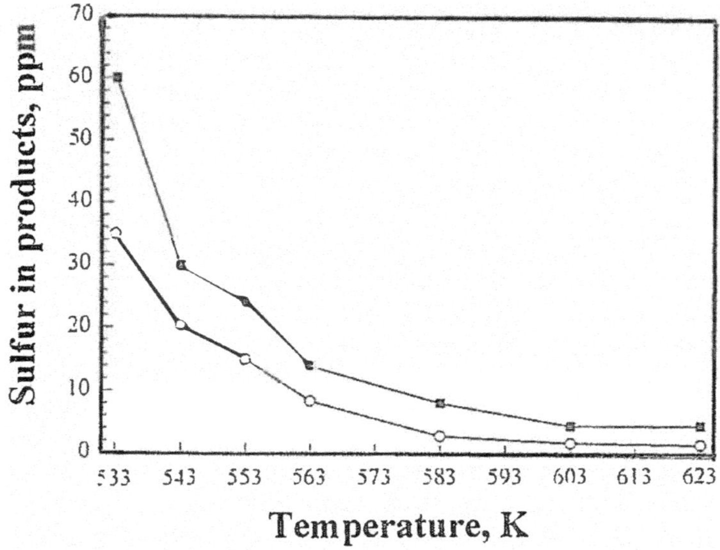

Figure 7.24 Effect of temperature on HDS over (■) Co–Mo/Al_2O_3 and (○) Co–Mo/CNT catalysts.
Reproduced from ref. 134 with permission from Elsevier, Copyright 2010.

temperature range. Thus, the lower concentration of sulfur remaining in the products confirmed the higher rate of HDS over the Co–Mo/CNT catalyst.

As a real feed, naphtha was used to evaluate the HDS activity of Co–Mo/MWCNT catalysts under near industrial conditions (*e.g.* trickle bed, 310 °C, 1.5 MPa, LHSV 4 h^{-1}).[135] Under these conditions, almost 90% sulfur removal was observed. The catalyst was prepared by incipient wetness co-impregnation of the CNTs using an aqueous solution of ammonium heptamolybdate and cobalt nitrate. The CNTs were prepared by the CVD method.

Gas oil derived from Athabasca bitumen was used as the feed to evaluate the activity of Ni–Mo catalysts supported on MWCNTs prepared by the CVD method using ferrocene as catalyst and toluene as carbon source.[136] The MWCNTs were functionalized by treatment with nitric acid before being impregnated with a solution of Mo precursor first followed by impregnation with a solution of Ni precursor. An Ni–Mo/Al_2O_3 catalyst containing the same amount of Ni and Mo was prepared in similar manner. The catalysts were sulfided with straight-run atmospheric gas oil spiked with 2.9% butanethiol. Furthermore, the catalysts were precooked with gas oil before being used in the HDS and HDN of the feed in a continuous fixed-bed reactor at 345–375 °C and 8 MPa. Under these conditions, the catalyst activity of the Ni–Mo/MWCNT catalyst was significantly higher than that of the conventional Ni–Mo/Al_2O_3 catalyst.

Sigurdson *et al.*[137] compared an Ni–Mo catalyst supported on MWCNTs with a commercial Ni–Mo–P/Al_2O_3 catalyst in the HDS and HDN of light gas oil obtained from Athabasca bitumen by coking. The comparison was performed on a volume basis. Thus, for both catalysts, 5 mL of catalyst was used. On a weight basis, this volume corresponded to 2 and 4 g for the Ni–Mo/CNT and Ni–Mo–P/Al_2O_3 catalysts, respectively. A test of 28 days' duration was performed in a continuous fixed-bed reactor (370 °C, 8.8 MPa, LHSV 2.0 h^{-1}). This was a study in which an Ni–Mo/CNT catalyst was evaluated under near commercial conditions. Both catalysts were subjected to a 3-day precoking stage to reach steady-state conditions. For the analysis of liquid products, the sampling was conducted at 24 h intervals. Figure 7.25[137] shows a steady performance of both catalysts during the 28-day trial, with average sulfur removal of 94.7 and 87.7% over the Ni–Mo–P/Al_2O_3 and Ni–Mo/CNT catalysts, respectively. Nitrogen removal averaged 58.9 and 89.6% over the Ni–Mo/CNT and Ni–Mo–P/Al_2O_3 catalysts, respectively. Again, for the latter, the amount of catalyst used for the test was twice that for the Ni–Mo/CNT catalyst. Therefore, on the basis of these results, it may be concluded that the Ni–Mo/CNT catalyst outperformed the commercial catalyst.

7.2.8 Summary of Studies of Conventional Metal Catalysts Supported on CNTs

Table 7.33 summarizes studies that were focused on the conversion of aromatic and heteroring-containing compounds and also real feeds derived

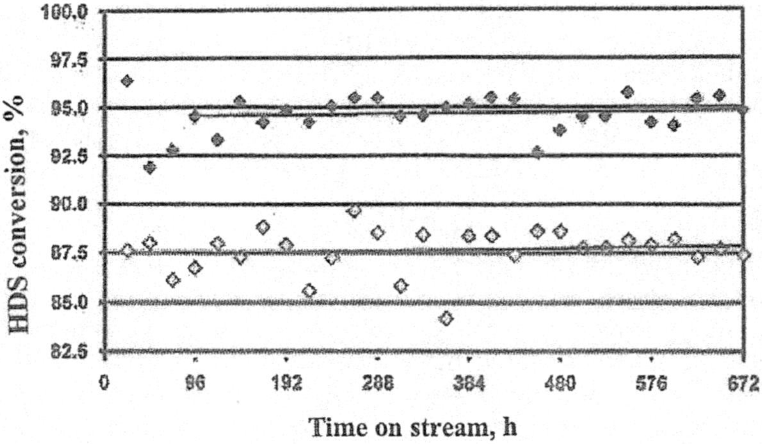

Figure 7.25 Steady-state activity in the HDS of gas oil: ♦, Ni–Mo/CNT; ◊, Ni–Mo–P/Al_2O_3.
Reproduced from ref. 137 with permission from John Wiley and Sons, Copyright © 2010 Canadian Society for Chemical Engineering.

Table 7.33 Conversions of petroleum-derived reactants and feeds over conventional metal catalysts supported on CNTs.

Catalyst	Reactants	Conditions	Ref.
Aromatics and heterorings			
1. Ni/CNT; Ni/AC	Benzene	433 K; 1 atm; continuous	119
2. Ni/CNT; Ni/AC	Benzene	433 K; 1 atm; continuous	120
3. NiP/CNT; Ni/CNT	Benzene	393–433 K; 1 atm; continuous	121
4. Ni/CNT; Mo/CNT; Ni–Mo/CNT	Thiophene	453–493; 1 atm; continuous; heptane	122
5. Ni–Mo–W/CNT	Thiophene	433 K, 1 atm; batch; decane	123
6. Co–Mo/CNT; Co–Mo/AC	Thiophene/ pyrrole	613 K; 1.5 MPa; continuous	124
7. Co–Mo/CNT	Thiophene	533 K; 1 atm; continuous	125
8. Co–Mo/CNT; Co–Mo/Al_2O_3	DBT	553 K; 2 MPa; continuous; toluene	126–128
9. Co–Mo/CNT	DBT	573 K; 3.5 MPa; batch; decalin	129
Real petroleum feeds			
10. Ni_2P/CNT	Naphtha	Continuous	130
11. Co–Mo/CNT; Co–Mo/Al_2O_3	Naphtha	583 K; 1.5 MPa; continuous	131, 132
12. Co–Mo/CNT; Co–Mo/Al_2O_3	Naphtha	533–623 K; 1.5 MPa; continuous	134
13. Co–Mo/CNT	Naphtha	583 K; 1.5 MPa; continuous	135
14. Ni–Mo/CNT; Ni–Mo/Al_2O_3	Gas oil	618–648 K; 8 MPa; continuous	136
15. Ni–Mo/CNT; Ni–Mo–P/CNT	Gas oil	643 K; 8.8 MPa; continuous	137
16. Ni/CNT; Ni/CB; Ni/zeolite	Heavy feed	573 K; 2–4 MPa; batch	133
17. Ni–Y/CNT	FCC feed	633 K; 1.5–2.5 MPa; continuous	138

from petroleum over transition metal catalysts supported on CNTs. Both batch and continuous systems were used for complete conversion tests. A comparison of the results in Table 7.31 with those in Table 7.33 shows that more severe conditions were necessary to achieve the conversion of petroleum-derived reactants and feeds to usable products than partial conversion of oxygenates. A comparison of the results in Tables 7.8 and 7.22 with those in Tables 7.31 and 7.33 indicates that for similar and/or the same reactants, more severe conditions were employed for conventional metals supported on CNTs than for noble metals supported on CNTs. However, more testing is necessary to confirm the difference. Nevertheless, the studies involving real feeds[133–138] showed that the CNT-supported catalysts outperformed the catalysts supported on ACs and CBs and also those supported on oxidic supports.

7.3 Rare Earth Metals Supported on CNTs

Yttrium (Y) is a rare earth metal rather than a noble metal, but with a price approaching that of the latter metals. In a few cases, Y was used for catalyst preparation. For example, Xu *et al.*[138] prepared bimetallic Ni–Y catalysts supported on SWCNTs and used them for the ultra-deep HDS of fluid catalytic cracking (FCC) gasoline and diesel. The preparation of the catalysts was conducted in either He or Ar to obtain Ni–Y/CNTHe and Ni–Y/CNTAr catalysts, respectively. In addition, the catalysts were preoxidized, *e.g.* to obtain Ni–Y/CNTHe-ox catalyst. The SWCNTs were prepared by an arc discharge method using a mixture of Ni and Y precursors with graphite powder. In this case, Ni and Y catalyzed SWCNT formation and ended up as nanosized particles as part of the catalyst. The bimetallic Ni–Y/SWCNT catalyst was more active than corresponding monometallic catalysts. The experiments were conducted in a continuous fixed-bed reactor at 250 and 360 °C and pressures of 1.5 and 2.5 MPa. Model gasoline and diesel consisting of thiophene and *n*-octane in addition to DBT and decahydronaphthalene, respectively, both containing 640 ppm of sulfur, were used as feeds in addition to real gasoline and diesel fractions from FCC containing 259 and 1126 ppm of sulfur, respectively. After the test, the sulfur content decreased to 1 and 10 ppm, respectively. Figure 7.26[138] shows the results for gasoline feed. It is evident that the type of catalyst had a significant effect on HDS.

Ma *et al.*[139] used rare earth metals such as cerium (Ce) and lanthanum (La) supported on CNTs as catalysts for the depolymerization of lignin dissolved in dioxane–methanol solvent. Before catalyst preparation, the raw CNTs were purified with nitric acid. After washing and drying, the catalysts were prepared *via* isometric impregnation using aqueous solutions of either Ce nitrate or La nitrate. Experiments were performed in an autoclave at 260 °C and 4 MPa of H_2 for 1 h. The Ce/CNT catalyst exhibited higher activity, giving more than 85% of good-quality fuel product. After four repeated cycles, the conversion of lignin over the Ce/CNT catalyst exceeded 80%.

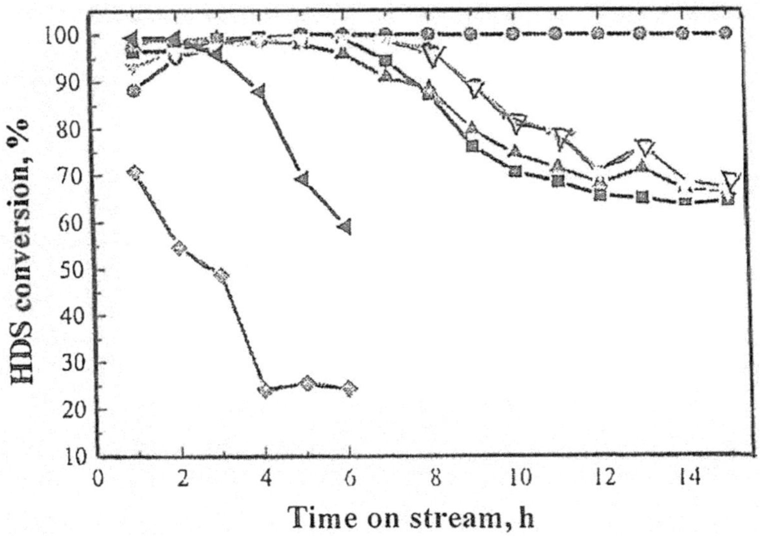

Figure 7.26 Effect of catalyst type on the HDS of gasoline: ■, Ni–Y/CNTHe; ●, Ni–Y/CNTHe-ox; △, Ni–Y/CNTAr; ▽, Ni–Y/CNTAr-ox; ◀, Ni/CNTHe; ◊, Y/CNTHe.
Reproduced from ref. 138 with permission from Elsevier, Copyright 2015.

References

1. S. H. Shuit, K. F. Yee, K. T. Lee, B. Subhash and S. H. Tan, *RSC Adv.*, 2013, **3**, 9070–9094.
2. X. Ni, B. Zhang, C. Li, M. Pang, D. Su, C. T. Williams and C. Liang, *Catal. Commun.*, 2012, **24**, 65–69.
3. Y. Wang, Z. Rong, Y. Wang, P. Zhang, Y. Wang and J. Qu, *J. Catal.*, 2015, **329**, 95–106.
4. H. Ma, T. Yu, X. Pan and X. Bao, *Chin. J. Catal.*, 2017, **38**, 1315–1321.
5. S. Takenaka, T. Iguchi, E. Tanabe, H. Matsune and M. Kishida, *Catal. Lett.*, 2011, **141**, 821–825.
6. Z.-T. Liu, C.-X. Wang, Z.-W. Liu and J. Lu, *Appl. Catal.*, 2008, **344**, 114–123.
7. J.-P. Tessonnier, L. Pesant, G. Ehret, M. J. Ledoux and C. Pham-Huu, *Appl. Catal.*, 2005, **288**, 203–210.
8. P. Bhanja, X. Liu and A. Modak, *ChemistrySelect*, 2017, **2**, 25.
9. Z. Tian, C. Liu, Q. Li, J. Hou, Y. Li and S. Ai, *Appl. Catal.*, 2015, **506**, 134–142.
10. S. M. George, *Chem. Rev.*, 2010, **110**, 111–131S.
11. X. Wang, Y. He, Y. Liu and X. Liang, *J. Catal.*, 2018, **366**, 61–69.
12. X. Wang, W. Ho, B. Deng and X. Liang, *J. Nanopart. Res.*, 2017, **19**, 153.
13. Y. Li, G. H. Lai and R. X. Zhou, *Appl. Surf. Sci.*, 2007, **253**, 4978–4984.
14. H. Ma, L. Wang, L. Chen, C. Dong, W. Yu, T. Huang and Y. Qian, *Catal. Commun.*, 2007, **8**, 452–456.

15. Y. Li, R. X. Zhou and G. H. Lai, *React. Kinet. Catal. Lett.*, 2006, **88**, 105–110.
16. J. Qiu, H. Zhang, X. Wang, H. Han, C. Liang and C. Li, *React. Kinet. Catal. Lett.*, 2006, **88**, 269–275.
17. Y. Li, Z.-G. Li and R.-X. Zhou, *J. Mol. Catal. A: Chem.*, 2008, **279**, 140–146.
18. H. Vu, F. Gonçalves, R. Philippe, E. Lamouroux, M. Corrias, Y. Kihn, D. Plee, P. Kalck and P. Serp, *J. Catal.*, 2006, **240**, 18–22.
19. Y. Li, C.-H. Ge, J. Zhao and R.-X. Zhou, *Catal. Lett.*, 2008, **126**, 280.
20. Y. Li, H. Cheng, W. Lin, C. Zhang, Q. Wu, F. Zhao and M. Arai, *Catal. Sci. Technol.*, 2018, **8**, 3580.
21. Y. Dai, X. Gao, X. Chu, C. Jiang, Y. Yao, Z. Guo, C. Zhou, C. Wang, H. Wang and Y. Yang, *J. Catal.*, 2018, **364**, 192–203.
22. W. Yan, Z. Guo, X. Jia, V. Kariwala and Y. Yang, *Chem. Eng. Sci.*, 2012, **76**, 26–36.
23. B. F. Machado, H. T. Gomes, P. Serp, P. Kalck and J. L. Faria, *ChemCatChem*, 2010, **2**, 190–197.
24. J. Amadou, K. Chizari, M. Houllé, I. Janowska, O. Ersen and E. Begin, *Catal. Today*, 2008, **138**, 62–68.
25. E. Asedegbega-Nieto, A. Guerrero-Ruiz and I. Rodríguez-Ramos, *Carbon*, 2006, **44**, 804–806.
26. X. Wang, W. Ho, B. Deng and X. Liang, *J. Nanopart. Res.*, 2017, **19**, 153.
27. X. Liang, Y. Zhou, J. Li and A. W. Weimer, *J. Nanopart. Res.*, 2011, **13**, 3781–3788.
28. X. Liang and C. Jiang, *J. Nanopart. Res.*, 2013, **15**, 1890.
29. G. Q. Guo, F. Qin, D. Yang, C. C. Wang, H. L. Xu and S. Yang, *Chem. Mater.*, 2008, **20**, 2291–2297.
30. S. A. Ananthan, N. Vengidusamy, K. Giribabu and R. Suresh, *Adv. Mater. Res.*, 2012, **584**, 229–233.
31. P. D. Zgolicz, J. P. Stassi, M. J. Yañez, O. A. Scelza and S. R. de Miguel, *J. Catal.*, 2012, **290**, 37–54.
32. F. Qin, W. Shen, C. Wang and H. Xu, *Catal. Commun.*, 2008, **9**, 2095–2098.
33. J. P. Stassi, V. I. Rodríguez, M. J. Yañez, S. R. de Miguel and P. D. Zgolicz, *Catal. Lett.*, 2017, **147**, 1903–1921.
34. V. Lordi, N. Yao and J. Wei, *Chem. Mater.*, 2001, **13**, 733–737.
35. H. Ren, W. Yu, M. Salciccioli, Y. Chen, Y. Huang, K. Xiong, D. G. Vlachos and J. G. Chen, *ChemSusChem*, 2013, **6**, 798–801.
36. L. S. Ribeiro, J. J. Delgado, J. J. de Melo, Órfão and M. F. R. Pereira, *Catal. Today*, 2017, **279**, 244–251.
37. L. S. Ribeiro, J. J. de Melo Orfao and M. F. Ribeiro Pereira, *Bioresour. Technol.*, 2017, **232**, 152–158.
38. L. S. Ribeiro, J. J. de Melo Órfão and M. F. Ribeiro Pereira, *Bioresour. Technol.*, 2017, **244**, 1173–1177.
39. (a) W. Deng, X. Tan, W. Fang, Q. Zhang and Y. Wang, *Catal. Lett.*, 2009, **133**, 167–174; (b) W. Deng, Y. Wang, Q. Zhang and Y. Wang, *Catal. Surv. Asia*, 2012, **16**, 91–105.

40. A. Aho, S. Roggan, K. Eränen, T. Salmia and D. Yu Murzin, *Catal. Sci. Technol.*, 2015, **5**, 953.
41. L. S. Ribeiro, J. J. Delgado, J. J. de Melo Órfão and M. F. Ribeiro Pereira, *ChemCatChem*, 2017, **9**, 888–896.
42. L. S. Ribeiro, J. J. Delgado, J. J. M. Órfão and M. F. R. Pereira, *Appl. Catal., B*, 2017, **217**, 265–274.
43. J. Pan, J. Li, C. Wang and Z. Yang, *React. Kinet. Catal. Lett.*, 2007, **90**, 233–242.
44. X. Guo, H. Dong, B. Li, L. Dong, X. Mua and X. Chen, *J. Mol. Catal. A: Chem.*, 2017, **426**, 79–87.
45. F. Sun, L. Chen, Y. Weng, T. Wang, S. Qiu, Q. Li, C. Wang, Q. Zhang and L. Ma, *Catal. Commun.*, 2017, **99**, 30–33.
46. R. F. Perez, O. S. G. P. Soares, A. M. Duarte de Farias, M. F. R. Pereira and M. A. Fraga, *Appl. Catal., B*, 2018, **232**, 101–107.
47. R. M. Mironenko, O. B. Belskaya, T. I. Gulyaeva, A. I. Nizovskii, A. V. Kalinkin, V. I. Bukhtiyarov, A. Lavrenov and V. A. Likholobov, *Catal. Today*, 2015, **249**, 145–152.
48. C. Wang, Z. Guo, Y. Yang, J. Chang and A. Borgna, *Ind. Eng. Chem. Res.*, 2014, **53**, 11284–11291.
49. L. Liu, H. Lou and M. Chen, *Appl. Catal.*, 2017, **550**, 1–10.
50. J. Kang, X. Liang and V. V. Guliants, *ChemCatChem*, 2017, **9**, 282–286.
51. P. Priecel, N. A. Endot, P. Demma Cara and J. A. Lopez, *Ind. Eng. Chem. Res.*, 2018, **57**, 1991–2002.
52. X. Wang, Y. Liu and X. Liang, *Green Chem.*, 2018, **20**, 2894–2902.
53. R. M. Mironenko, O. B. Belskaya, A. V. Lavrenova and V. A. Likholobova, *Kin. Catal.*, 2018, **59**, 339–346.
54. Y. Liu, Z. Chen, X. Wang, Y. Liang, X. Yang and Z. Wang, *ACS Sustainable Chem. Eng.*, 2017, **5**, 744–751.
55. A. Piskun, J. G. M. Winkelman, Z. Tang and H. J. Heeres, *Catalysts*, 2016, **6**, 131–150.
56. X. Du, Y. Liu, J. Wang, Y. Cao and K. Fan, *Chin. J. Catal.*, 2013, **34**, 993–1001.
57. K. B. Jung, J. Lee, J.-M. Ha, H. Lee, D. J. Suh, C.-H. Jun and J. Jae, *Catal. Today*, 2018, **303**, 191–199.
58. M.-Y. Chen, Y.-B. Huang, H. Pang, X.-X. Liu and Y. Fu, *Green Chem.*, 2015, **17**, 1710–1717.
59. X. Yang, Y. Liang, Y. Cheng, W. Song, X. Wang, Z. Wang and J. Qiu, *Catal. Commun.*, 2014, **47**, 28–31.
60. Q. Wang, G. Neeraj, G. Wen, A. H. Sharifah and D. S. Su, *J. Energy Chem.*, 2017, **26**, 8–16.
61. A. A. Dwiatmoko, L. Zhou, I. Kim, J.-W. Choi, D. J. Suh and J.-M. Ha, *Catal. Today*, 2016, **265**, 192–198.
62. S. L. Yohe, H. J. Choudhari, D. D. Mehta, P. J. Dietrich, M. D. Detwiler, C. M. Akatay, E. A. Stach, J. T. Miller, W. N. Delgass, R. Agrawal and F. H. Ribeiro, *J. Catal.*, 2016, **344**, 535–552.

63. H. Ohta, H. Kobayashi, K. Hara and A. Fukuoka, *Chem. Commun.*, 2011, **47**, 12209–12211.
64. F. Li, B. Cao, W. Zhu, H. Song, K. Wang and C. Li, *Catalysts*, 2017, 7, 145.
65. Y. Xiang, L. Kong, P. Xie, T. Xu, J. Wang and X. Li, *Ind. Eng. Chem. Res.*, 2014, **53**, 2197–2203.
66. L. Zhang, B. Wang, Y. Ding, G. Wen, S. B. Abd Hamid and D. Su, *Catal. Sci. Technol.*, 2016, **6**, 1003.
67. C. Yang, R. Nie, J. Fu, Z. Hou and X. Lu, *Bioresour. Technol.*, 2013, **146**, 569–573.
68. J. Li, S. Wang, H.-Y. Liu, H.-J. Zhou and Y. Fu, *ChemistrySelect*, 2017, **2**, 33–41.
69. Y. Wang, X. Yang, Y. Chen, S. Nie and M. Xie, *React. Kinet. Catal. Lett.*, 2017, **122**, 915–930.
70. Y.-Z. Xiang, Y.-A. Lv, T.-Y. Xu, X.-N. Li and J.-G. Wang, *J. Mol. Catal. A: Chem.*, 2011, **351**, 70–75.
71. Y. Zhou, J. Liu, X. Li, X. Pan and X. Bao, *J. Nat. Gas Chem.*, 2012, **21**, 241–245.
72. Y. Huang, Y. Ma, Y. Cheng, L. Wang and X. Li, *Ind. Eng. Chem. Res.*, 2014, **53**, 4604–4613.
73. S. Tourani, A. M. Rashidi, A. A. Safekordi, H. R. Aghabozorg and F. Khorasheh, *Ind. Eng. Chem. Res.*, 2015, **54**, 7591–7603.
74. S. Tourani, A. A. Safekordi, M. Rashidzadeh, A. M. Rashidi and F. Khorashehe, *Chem. Eng. Res. Des.*, 2016, **109**, 41–52.
75. S. Tourani, F. Khorasheh, A. M. Rashidi and A. A. Safekordi, *J. Ind. Eng. Chem.*, 2015, **28**, 202–210.
76. E. Furimsky, *Catal. Today*, 2013, **217**, 13–56.
77. E. Furimsky, *Appl. Catal.*, 2000, **199**, 147–188.
78. E. N. Terekhova, O. N. Baklanova and A. V. Lavrenov, *Catal. Ind.*, 2017, **9**, 110–121.
79. B. Yoon, H.-B. Pan and C. M. Wai, *J. Phys. Chem. C*, 2009, **113**, 1520–1525.
80. Z. Y. Sun, Z. M. Liu, B. X. Han, Y. Wang, J. M. Du, Z. L. Xie and G. J. Han, *Adv. Mater.*, 2005, **17**, 928.
81. H.-B. Pan and C. M. Wai, *New J. Chem.*, 2011, **35**, 1649–1660.
82. H. Kong, M. Zhou, G.-D. Lin and H.-B. Zhang, *Catal. Lett.*, 2010, **135**, 83–90.
83. A. M. Zhang, J. L. Dong, Q. H. Xu, H. K. Rhee and X. L. Li, *Catal. Today*, 2004, **93/95**, 347–352.
84. B. Pawelec, V. La Parola, R. M. Navarro, S. Murcia-Mascarós and J. L. G. Fierro, *Carbon*, 2006, **44**, 84–96.
85. M. Zhou, G. Lin and H. Zhang, *Chin. J. Catal.*, 2007, **28**, 210–216.
86. Y. Ma, Y. Huang, Y. Cheng, L. Wang and X. Li, *Appl. Catal.*, 2014, **484**, 154–160.
87. H. Bazzazzadegan, M. Kazemeini and A. M. Rashidi, *Res. Chem. Intermed.*, 2015, **41**, 1023–1034.

88. L. Shao, B. Zhang, W. Zhang, D. Teschner, F. Girgsdies, R. Schlögl and D. S. Su, *Chem. - Europ. J.*, 2012, **18**, 14962–14966.
89. S. Domínguez-Domínguez, A. Bereguera-Murcia, B. K. Pradham, A. Linares-Solano and D. Cazorla-Amorós, *J. Phys. Chem. C*, 2008, **112**, 3827–3834.
90. E. V. Starodubtseva, M. G. Vinogradov, O. V. Turova, N. A. Bumagin, E. G. Rakov and V. I. Sokolov, *Catal. Commun.*, 2009, **10**, 1441–1442.
91. P. Liu, W.-T. Chang, M.-Y. Wu, Y.-X. Li and J. Wang, *React. Kinet., Mech. Catal.*, 2015, **116**, 409–419.
92. C. Li, Z. Shao, M. Pang, C. T. Williams, X. Zhang and C. Liang, *Ind. Eng. Chem. Res.*, 2012, **51**, 4934–4941.
93. P. Chen, L.-M. Chew, A. Kostka, M. Muhler and W. Xia, *Catal. Sci. Technol.*, 2013, **3**, 1964–1972.
94. M. Cano, A. M. Benito, W. K. Masera and E. P. Urriolabeitia, *New J. Chem.*, 2013, **37**, 1968.
95. L. Vanyorek, E. Sikora, A. Kiss, A. Sike, Z. G. Hutkai, P. Pekker, B. Viskolcz and B. Fiser, *React. Kinet. Mech. Catal.*, 2018, **125**, 37–46.
96. Y. Xu, L. Chen and X. Wang, *J. Chem.*, 2013, **91**, 307–314.
97. L.-P. Xiao, S. Wang, H. Li, Z. Li, Z.-J. Shi, L. Xiao, R.-C. Sun, Y. Fang and G. Gong, *ACS Catal.*, 2017, 7, 7535–7542.
98. M. Zhou, G. Xiao, K. Wang and J. Jiang, *Fuel*, 2016, **180**, 749–758.
99. A. B. Dongil, L. Pastor-Pérez, A. Sepúlveda-Escribano, R. García and N. Escalona, *Fuel*, 2016, **172**, 65–69.
100. A. B. Dongil, I. T. Ghampson, R. García, J. L. G. Fierro and N. Escalona, *RSC Adv.*, 2016, **6**, 2611–2623.
101. A. B. Dongil, B. Bachiller-Baeza, I. Rodriguez-Ramos, J. L. G. Fierro and N. Escalona, *RSC Adv.*, 2016, **6**, 26658–26667.
102. H. Fang, J. Zheng, X. Luo, J. Du, A. Roldan, S. Leoni and Y. Yuan, *Appl. Catal.*, 2017, **529**, 20–31.
103. B. Rahzani, M. Saidi, H. R. Rahimpour, B. C. Gates and M. R. Rahimpour, *RSC Adv.*, 2017, 7, 10545–10556.
104. L. Yu, L. He, J. Chen, J. Zheng, L. Ye, H. Lin and Y. Yuan, *ChemCatChem*, 2015, 7, 1701–1707.
105. L. J. Liu, H. Lou and M. Chen, *Int. J. Hydrogen Energy*, 2016, **41**, 14721–14731.
106. (a) M. Zhou, L. Tian, L. Niu, C. Li, G. Xiao and R. Xiao, *Fuel Process. Technol.*, 2014, **126**, 12–18; (b) W. Gong, C. Chen, H. Zhang, G. Wang and H. Zhao, *Catal. Sci. Technol.*, 2018, **8**, 5506–5514.
107. E. F. Mai, M. A. Machado, T. E. Davies, J. A. Lopez-Sanchez and V. Teixeira da Silva, *Green Chem.*, 2014, **16**, 4092–4097.
108. J. Quiroz, E. F. Mai and V. Teixeira da Silva, *Top. Catal.*, 2016, **59**, 148–158.
109. J. M. Liang, R. R. Ding, Y. L. Wu, Y. Chen, K. J. Wu, Y. Q. Meng, M. D. Yang and Y. W. Wang, *J. Mol. Catal. A: Chem.*, 2016, **411**, 95–102.

110. J. Han, J. Duan, P. Chen, H. Lou, X. Zheng and H. Hong, *Green Chem.*, 2011, **13**, 2561–2568.
111. R. R. Ding, Y. L. Wu, Y. Chen, J. M. Liang, J. Liu and M. D. Yang, *Chem. Eng. Sci.*, 2015, **135**, 517–525.
112. R. T. Ding, Y. L. Wu, Y. Chen, H. Chen, J. L. Wang, Y. C. Shi and M. D. Yang, *Catal. Sci. Technol.*, 2016, **6**, 2065–2076.
113. Y. Chen, R. Mu, M. Yang, L. Fang and J. Gong, *Chem. Eng. Sci.*, 2017, **161**, 299–307.
114. Y. Duan, R. Ding, Y. Shi, X. Fang, H. Hu, M. Yang and Y. Wu, *Catalysts*, 2017, **7**, 81.
115. L. J. Malobela, J. Heveling, W. G. Augustyn and L. M. Cele, *Ind. Eng. Chem. Res.*, 2014, **53**, 13910–13919.
116. H. N. Hareesh, K. U. Minchitha, K. Venkatesh, N. Nagaraju and N. Kathyayini, *RSC Adv.*, 2016, **6**, 82359–82369.
117. Y. Tang, D. Yang, F. Qin, J. Hu, C. Wang and H. Xu, *J. Solid State Chem.*, 2009, **182**, 2279–2284.
118. L. Zhang, X. Chen, S. Jin, J. Guan, C. T. Williams, Z. Peng and C. Liang, *J. Solid State Chem.*, 2014, **217**, 105–112.
119. Y. Sun, C. Li and A. Zhang, *Appl. Catal.*, 2016, **522**, 180–187.
120. H. Yang, S. Song, R. Rao, X. Wang, Q. Yu and A. Zhang, *J. Mol. Catal. A: Chem.*, 2010, **323**, 33–39.
121. Y. Ju and F. Li, *J. Nat. Gas Chem.*, 2006, **15**, 313–318.
122. S. Liu, Q. Jin, Y. Xu, X. Fang, N. Liu, J. Zhang, X. Liang and B. Chen, *Fuel*, 2018, **232**, 36–44.
123. D. M. Nejad, N. Rahemi and S. Allahyari, *React. Kinet., Mech. Catal.*, 2017, **120**, 279–294.
124. K. Dong, X. Ma, H. Zhang and G. Lin, *J. Nat. Gas Chem.*, 2010, **15**, 28–37.
125. J. Zhang, W. Yin, H. Shang and C. Liu, *J. Nat. Gas Chem.*, 2008, **17**, 165–170.
126. H. Shang, C. Liu, Y. Xu, J. Qiu and F. Wei, *Fuel Process. Technol.*, 2007, **88**, 117–123.
127. H.-Y. Shang, C.-G. Liu, R.-Y. Zhao, M.-B. Wu and F. Wei, *Chin. J. Chem.*, 2004, **22**, 1250–1256.
128. H. Shang, C. Liu, Y. Xu, J. Qiu and F. Wei, *J. Nat. Gas Chem.*, 2006, **15**, 203–210.
129. J. Whelan, M. S. Katsiotis, S. Stephen, G. E. Luckachan, A. Tharalekshmy, N. D. Banu, J.-C. Idrobo, S. T. Pantelides, R. V. Vladea, I. Banu and S. M. Alhassan, *Energy Fuels*, 2018, **32**, 7820–7826.
130. M. Kiani, H. Aghabozorg, K. J. Jozani, A. Rashidi and M. Mohsennia, *Phosphorus, Sulfur Silicon Relat. Elem.*, 2013, **188**, 1254–1261.
131. E. Soghrati, M. Kazemeini, A. M. Rashidi and K. J. Jozani, *Procedia Eng.*, 2012, **292/42**, 1484–1492.
132. E. Soghrati, M. Kazemeini, A. M. Rashidi and K. J. Jozani, *J. Taiwan Inst. Chem. Eng.*, 2014, **54**, 887–895.
133. K. Guo, Y. Zhang, Q. Shi and Z. Yu, *Energy Fuels*, 2017, **31**, 6045–6055.

134. A. Nakhaei Pour, K. Morad Rashidi, A. Jafari Jozani, A. Mohajeri and P. Khorami, *J. Nat. Gas Chem.*, 2010, **19**, 91–95.
135. P. Khorami and M. Kalbasi, *J. Exp. Nanosci.*, 2011, **6**, 349–361.
136. I. Eswaramoorthi, V. Sundaramurthy, Nikhil Das, A. K. Dalai and J. Adjay, *Appl. Catal.*, 2008, **339**, 187–195.
137. S. Sigurdson, A. K. Dalai and J. Adjaye, *Can. J. Chem. Eng.*, 2011, **89**, 562–575.
138. K. Xu, Y. Li, X. Xu, C. Zhou, Z. Liu, F. Yang, G. Wang, J. Gao and C. Xu, *Fuel*, 2015, **160**, 291–296.
139. Q. Ma, Q. Liu, W. Li, L. Ma, J. Wang, M. Liu and Q. Zhang, *Fuel Process. Technol.*, 2017, **161**, 220–225.

CHAPTER 8

Catalysts Supported on Carbon Nanofibers and Carbon Nanohorns

Interest in CNF- and CNH-supported catalysts has been noted, although to a lesser extent than CNT-supported catalysts; CNHs have attracted the least attention. The combination of a large pore volume (*e.g.* 0.5–2 $cm^3 g^{-1}$), an open morphology and a suitable surface area (*e.g.* 100–200 $m^2 g^{-1}$) suggests that CNFs have potential for use as catalyst supports.[1,2] Moreover, the predominance of the edge regions in CNFs (Figure 2.4) compared with the basal planes in CNTs (Figure 2.3) suggests that anchoring of active metals on the former can be achieved more readily. Both noble and conventional metals have been used as active metals. In some respects, similarities between CNF and CNT supports may be evident. For example, similar methods can be used for the functionalization and doping of CNFs and also for catalyst preparation. Furthermore, the same noble and conventional metals have been used for catalyst preparation. Biomass-derived feeds and reactants and also those derived from petroleum that were used over CNT-supported catalysts have been the focus of attention in studies on CNF-supported catalysts.

Similarly to CNTs, the hydrophobicity of CNFs requires attention, particularly during catalyst preparation. Several procedures have been developed to alleviate this problem using functionalization methods[3–11] and doping[12–17] of carbon nanomaterials. For example, Kwande *et al.*[18] showed that for CNFs, the hydrophobicity problems can be overcome by using colloidal methods of preparation. Thus, *via* a modified polyol method, up to 24 wt% of Pt particles (2–4 nm) could be deposited. Much smaller amounts

Catalysis Series No. 37
Carbon Nanomaterials in Hydrogenation Catalysis
By Edward Furimsky

Published by the Royal Society of Chemistry, www.rsc.org

of Pt could be deposited by the widely used incipient wetness impregnation method.

Studies in which CNF-supported catalysts were evaluated in parallel with catalysts supported on other carbon nanosupports, amorphous carbons and oxidic supports and also with other types of catalysts have been noted and are of great importance for catalyst development and catalyst performance comparisons. Such results are important particularly if the testing conditions and reactants were similar.

8.1 Noble Metals Supported on CNFs

For the development and testing of noble metal catalysts supported on CNFs, oxygen-containing reactants such as α,β-unsaturated aldehydes, particularly CALD and citral, have frequently been used in addition to reactants derived from lignocellulosic biomass. Attention has been paid to unsaturated hydrocarbons and aromatics and other petroleum-derived reactants. Chlorinated hydrocarbons have been used to a lesser extent.

8.1.1 Conversion of α,β-Unsaturated Aldehydes

One of the most frequently studied model reactions, the HYD of CALD (Figure 5.2), was investigated by Plomp *et al.*[19] over Pt/CNF and Ru/CNF catalysts prepared by two different methods. First, in method A, Ru was deposited on CNF as part of an aqueous suspension containing an Ru precursor such as $RuNO(NO_3)_3 \cdot xH_2O$ and urea with vigorous stirring. The solid was filtered off, washed, dried and reduced in H_2 at 473 K. Second, in method B, CNFs were contacted with either $Ru_3(CO)_{12}$ or $Pt(acac)_2$ in a flow of N_2 at 413 and 453 K, respectively. The catalysts were then reduced in a flow of H_2–N_2. Selected catalysts were thermally treated at 973 K to remove the excess of O-containing groups formed during the CNF functionalization treatment. The size of the metal particles on the catalysts prepared by methods A and B differed significantly, *i.e.* 1–3 and 2–6 nm, respectively. The CALD conversions over these catalysts, obtained at 313 K and near atmospheric pressure H_2, are shown in Figure 8.1. Without thermal pretreatment, over both Pt/CNF and Ru/CNF catalysts, the CALD conversion approached about 25%. For thermally treated catalysts (*e.g.* Pt/CNF-973), a dramatic increase in conversion was observed. The catalyst prepared by method A was more active. Initially, this catalyst exhibited high selectivity for the targeted product such as CALC. On the other hand, the heat-treated catalyst prepared by method B was more selective with respect to HCALD. The selectivity difference was attributed to the different sizes of the metal particles on the catalysts. Thus, the smaller sized particles on the Pt/CNF catalyst facilitated the interaction of CALD with Pt metal *via* the C=O bond rather than *via* the C=C bond. This suggests that the catalyst selectivity can be optimized by modifying the conditions used during catalyst preparation.

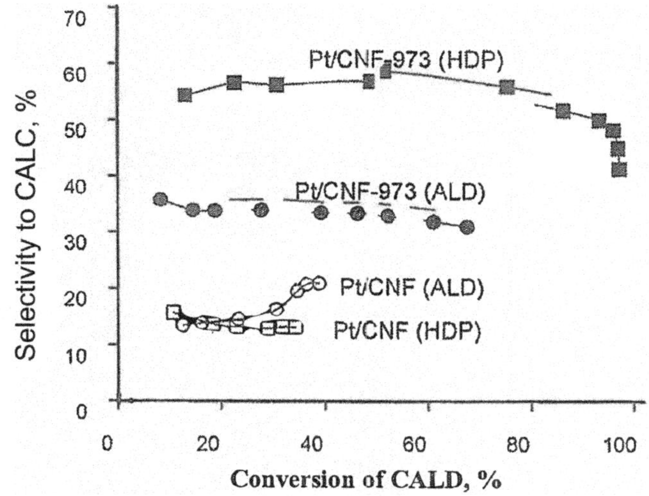

Figure 8.1 Effect of Pt/CNF catalyst pretreatment on selectivity in CALD conversion: ■, heat treated (method A); ●, heat treated (method B); □, untreated (method A); ○, untreated (method B).
Reproduced from ref. 19 with permission from Elsevier, Copyright 2008.

CNFs with mean diameters of 40–50 nm and lengths of more than 10 μm were used as supports for Pd catalysts.[20,21] The CNFs were prepared by the decomposition of an ethane–H_2 mixture over Ni/Al_2O_3 catalyst at 650 °C. A Pd(5%)/CNF catalyst was prepared by incipient wetness impregnation of nitric acid-treated CNFs using an aqueous solution of Pd nitrate. Subsequently, the vacuum-dried solid was reduced at 350 °C in a flow of H_2 for 2 h. The HYD of CALD in dioxane was carried out at 353 K and near atmospheric pressure H_2 in a continuous system. At 353 K, the catalyst exhibited high selectivity for the HYD of the C=C bond, leading to the formation of HCALD. Under the same conditions, the activity and selectivity of the Pd/CNF catalysts were significantly greater than those of a commercial Pd/AC catalyst.

Jung *et al.*[22] compared functionalized CNTs with platelet CNFs (P-CNFs) and herringbone CNFs (H-CNFs) as supports for Pt catalysts. Several methods were used for the metal loading, *i.e.* conventional wetness impregnation (WI), ion adsorption (IA), homogeneous deposition–precipitation (HDP) and colloidal microwave process (CMP) methods. The loading *via* the CMP method yielded the most active catalysts. The conventional AC, loaded with the CMP method, and commercial Pt/AC and Pt/Al_2O_3 catalysts were used for comparison. These catalysts were applied in the HYD of CALD to CALC (batch process, 383 K, 5 MPa H_2). Figure 8.2 shows the correlation between conversion and selectivity. It is evident that the Pt/H-CNF catalyst was more selective with respect to CALC than the corresponding Pt/MWCNT catalyst. Thus, at 80% conversion, the selectivity of the former approached 70% compared with about 40% for the Pt/MWCNT catalyst. The commercial Pt/AC and Pt/Al_2O_3 catalysts exhibited much lower selectivity.

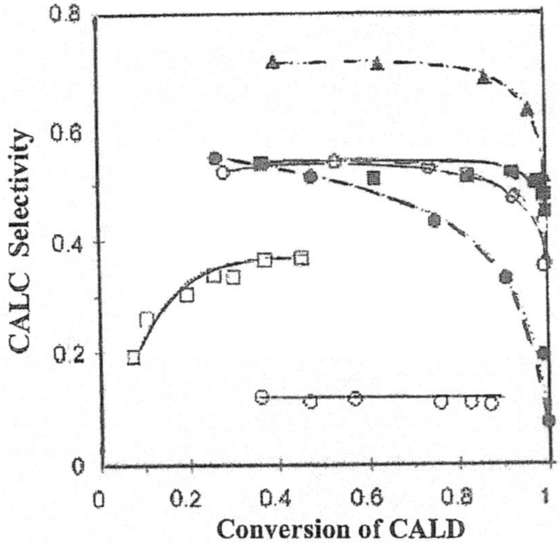

Figure 8.2 Effect of support type on the selectivity of Pt catalysts: ▲, Pt/H-CNF; ■, Pt/P-CNF; ●, Pt/CNT; ☼, Pt/AC1; ○, Pt/AC2; □, Pt/Al_2O_3.[22]
Reproduced from ref. 22 with permission from Elsevier, Copyright 2009.

Table 8.1 Selectivity in CALD conversion over Pt/CNF catalysts at 60% conversion (383 K, 4.8 MPa H_2).[23]

	Selectivity/mol%			By-products[a]	
Catalyst	CALC	HCALD	HCALC	A	B
Pt/CNFnp[b]	58	22	13	1	1
Pt/CNF	37	32	15	2	6
Pt/CNF573	13	23	12	11	19
Pt/CNF773	13	61	17	1	2
Pt/CNF973	8	74	18	0	0

[a]A, cinnamyl isopropyl ether; B, diisopropylacetal of hydrocinnamaldehyde.
[b]np, not prereduced.

Toebes *et al.*[23] studied the HYD of CALD over a series of Pt/CNF catalysts in a continuous system at 383 K and 4.8 MPa. For the tests, CALD was dissolved in 2-propanol. The Pt/CNF catalysts were prepared by vigorously stirring an acidified suspension of CNFs with $Pt(NH_3)_4(NO_3)_2$. After filtering, washing and drying, the solid was reduced in H_2 at 473 K. To obtain catalysts with varying content of O-containing groups, the catalyst was heat treated in N_2 for 2 h at 573, 773 and 973 K. The results for selectivity over these catalysts are given in Table 8.1.[23] It is evident that the catalyst pretreatment, *i.e.* either prereduction or thermal treatment, had a significant effect on selectivity. This confirmed the involvement of O-containing groups because their concentration on the CNF surface was influenced by the pretreatment. Similar effects of O-containing groups were observed in the HYD of CALD over an Ru/CNF catalyst.[24]

The selective HYD of CALD over Pd/CNF, Pd/CNT and a commercial Pd/AC (all containing 5% Pd) was conducted by Ledoux *et al.*[25] at 80 °C and with bubbling of H_2 using CALD dissolved in dioxane. The catalysts were prepared by incipient wetness impregnation of the nitric acid-treated supports using an aqueous solution of a Pd precursor. The selectivity was determined as the yield of HCALC of the converted CALD. It is noted that only traces of CALC were formed under these conditions. The performances of the catalysts are compared in Figure 8.3. The selectivity of the Pd/CNF and Pd/CNT catalysts with respect to HCALC was fairly high compared with that of the Pd/AC catalyst. Also, the activity of the Pd/AC catalyst was very low. Thus, for this catalyst, the time required for complete conversion of CALD was twice that observed over the Pd/CNT and Pd/CNF catalysts. Based on the results for Pd/CNT and PD/CNF in Figure 8.3, Pd/CNT exhibited higher activity than Pd/CNF. Thus, complete conversion of CALD was achieved in less than 25 min whereas over the latter catalyst more than 30 min were required. Both catalysts exhibited high selectivity for HYD of the C=C bond yielding HCALD.

Zhu and co-workers[26–28] prepared a CNF–TiO_2 composite support by the decomposition of methane catalyzed by Ni–Cu. The preparation of the CNF–TiO_2 composite began with the impregnation of TiO_2 cylinders with Ni and Cu nitrates. After drying and calcining at 773 K, the solid was reduced at 800 K in a flow of N_2–H_2. The carbon component contained 95% of CNFs and 5% of amorphous carbon. The latter together with the remaining Ni and Cu metals were removed by nitric acid treatment. Pd/CNF–TiO_2 catalyst was prepared by impregnation of the CNF–TiO_2 support using Pd–ethylenediaminetetraacetic acid (EDTA) chelates. The catalyst was reduced at 363 K with hydrazine hydrate. The conversion of citral to citronellal was studied in an autoclave at 363 K and 2 MPa of H_2. Under these conditions, the selectivity with respect to citronellal approached 88% at 90% conversion of citral. With this performance, the Pd/CNF–TiO_2 catalyst was superior to a Pd/AC catalyst in spite of the much greater surface area of the latter. However, most of the surface area of the Pd/AC was in the micropore

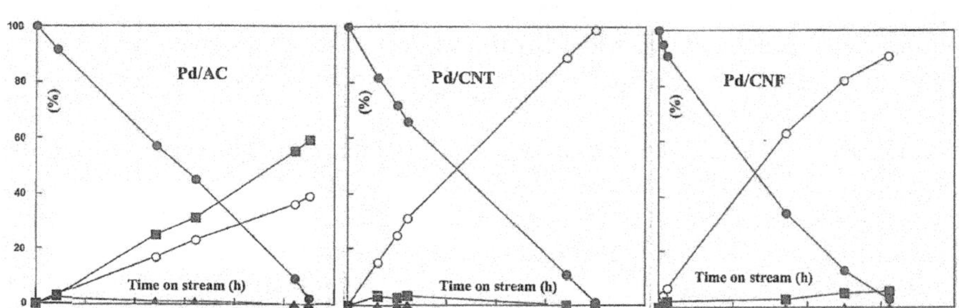

Figure 8.3 Effect of support type on the product distribution in CALD conversion: ●, conversion; ■, HCALC; ○, HCALD; ▲, CALC.
Reproduced from ref. 25 with permission from Elsevier, Copyright 2003.

range whereas that for the Pd/CNF–TiO_2 was in the mesopore range. Hence, for the latter catalyst, diffusion limitations should be much less evident.

In a study by Aumo *et al.*,[29] Pt/CNF catalyst was compared with a commercial Pt/AC catalyst in the HYD of citral (Figure 5.3) in a batch reactor at 80 and 100 °C and 0.5 and 1.0 MPa of H_2. The Pt/CNF catalyst exhibited high selectivity for the HYD of the C=O bond, giving geraniol and nerol as the main products, whereas over the commercial Pt/AC catalyst the formation of citronellal in addition to other products was favored. A geraniol : nerol ratio of 2.3 was maintained during the entire experiment, hence it was not influenced by either citral conversion or catalyst deactivation.

8.1.2 Lignocellulosic Reactants

These reactants can be identified in the mechanism of the depolymerization of cellulose and hemicellulose (Figure 5.1), which are important components of lignocellulosic biomass, in addition to lignin (Figure 5.5). They include both the final products and intermediates shown in this mechanism. These reactants require further upgrading to useful products.

Frequently, catalysts exhibiting a suitable selectivity are necessary. Some of these reactants and feeds were tested over CNT-supported catalysts.

Unique methods of catalyst preparation and testing were employed by Zhou *et al.*[30] An Ru catalyst supported on a CNF–graphite (CNF–GF) composite was used for the conversion of an aqueous solution of sorbitol in an autoclave equipped with a stirrer. For the experiments conducted at 220 °C and 8 MPa of H_2, 2×1 cm rectangles of catalyst material were mounted on the stirrer as stirring blades. The catalyst preparation began with cutting the graphite into predefined shapes, which were subsequently impregnated using an ethanolic solution of Ni nitrate. The CNFs were then grown on the graphite using ethane as the carbon source. After washing with HCl, the composites were treated with either H_2O_2 or nitric acid. Ru(3%) catalysts were prepared by incipient wetness impregnation of the CNF–GF composite using an aqueous solution of $RuCl_3$. Before testing, the structured catalyst was aged, dried and reduced in a flow of H_2–Ar. In the HYD of sorbitol, the activity and selectivity of the catalysts with respect to ethylene glycol, propylene glycol and glycerol were compared with the results for a powdered Ru(3%)/CNF catalyst. The structured catalyst was less active but its selectivity approached 80%.

Nitric acid-purified CNFs were used by Zhou *et al.*[31] as the support for the preparation of Ru/CNF catalysts by incipient wetness impregnation using $RuCl_3{\cdot}H_2O$ as the Ru precursor. The catalyst was used for the conversion of sorbitol to glycols in a batch reactor at 220 °C and 8 MPa of H_2 for 4 h. The experiments were performed in the presence of basic oxides such as NaOH, KOH, $Mg(OH)_2$, $Ba(OH)_2$ and CaO. The bases enhanced the conversion of sorbitol but had different effects on the product distribution. The highest yields of glycol were obtained in the presence of CaO. Apparently, the CaO facilitated the optimal basicity required for hydrogenolysis reactions.

Table 8.2 Activity and selectivity of Ru/CNF and Ru/AC catalysts in the conversion of sorbitol.[32]

		Selectivity[b]/%			
Catalyst[a]	Conversion/%	A	B	C	D
Ru/CNF	85.7	9.5	19.3	32.0	51.3
Ru/AC	71.4	1.5	10.5	15.6	26.1

[a]Conditions: 1.5 g catalyst, 10.0 g CaO, 220 °C, 2 MPa H_2, 4 h.
[b]A, glycerol; B, ethylene glycol; C, propylene glycol, D, sum of glycols.

Zhao *et al.*[32] prepared a CNF support by a catalytic (Fe oxide) CVD method *via* decomposition of CO in a flow of CO–H_2 (4 : 1) at 600 °C. The CNFs had a narrow size distribution centered at around 1–2 nm. The CNFs were purified with HCl solution before being impregnated with $RuCl_3$ solution by the incipient wetness impregnation method. A commercial Ru/AC catalyst was used for comparison. After activation, the catalysts were used for the conversion of sorbitol in an autoclave at 220 °C and 2 MPa of H_2 for 4 h. The results of these tests are given in Table 8.2.[32] It is evident that in terms of conversion to desirable products such as glycols, the Ru/CNF catalyst outperformed the Ru/AC catalyst.

Yang *et al.*[33] applied a novel approach to prepare Ru nanoparticle-inserted CNFs *via* the reaction of $RuCl_3$ and Zn acetate with trimesic acid [benzene-1,3,5-tricarboxylic (H3BTC) acid] in the presence of *N*,*N*-dimethylformamide. The resulting precipitate was washed and vacuum dried to obtain Ru–ZnBTC fibers, which were subsequently thermally treated in N_2 at either 750 or 950 °C to obtain RuZn/CNF and Ru/CNF-950, respectively. The RuZn/CNF was further treated with HCl to remove Zn to obtain Ru/CNF-750. The catalysts were used for the HYD of LA to GVL. The tests were conducted in an autoclave at 159 °C and 4.5 MPa of H_2 for 8 h. Under these conditions, over Ru(0.27%)/CNF-950, the conversion of LA and the selectivity with respect to GVL were 96 and 95%, respectively, compared with 41% over Ru(0.21%)/CNF-750. Over a commercial Ru(5%)/AC catalyst, the conversion and selectivity approached 98 and 95%, respectively. However, on the basis of a unit of Ru, the reaction rate over both Ru/CNF catalysts was more than an order of magnitude greater than that over the Ru/AC catalyst.

8.1.3 Conversion of Hydrocarbons

Alkanes, alkenes and alkynes have been focus of attention mainly as part of HYD and hydroisomerization (HIS) reactions. Little attention has been paid to the HYD of aromatics over CNF-supported catalysts.

8.1.3.1 Alkanes

Pt(5%)/CNF catalyst exhibited high activity and selectivity in the HIS of *n*-hexane to isoalkanes.[34] The catalyst was prepared by impregnation of platelet-type CNFs (Figure 2.4) using a Pt precursor such as $[Pt(NH_3)_2](NO_3)_2$ dissolved in *n*-butanol. The solid was dried at 323 K, calcined at 523 K and reduced in H_2 at 573 K. The tests were conducted in a closed-loop circulating system using a mixture of *n*-hexane and H_2 at 483 and 603 K. The selectivity with respect to isoalkanes exceeded 60% with benzene and cracking products accounting for the rest; *n*-hexenes were also present but only in trace amounts.

8.1.3.2 Alkynes

Chesnokov *et al.*[35] prepared CNFs by methane decomposition on an Ni–Cu–Fe/Al_2O_3 catalyst. After purification, the CNFs were used as supports for monometallic Pd and bimetallic Pd–Ag catalysts. The amount of Pd varied from 0 to 0.3 wt%. The catalysts were compared in the selective HYD of acetylene to ethylene in a continuous-flow reactor at 35–150 °C. The conversion of acetylene to ethylene increased with increase in temperature, *i.e.* at 40 and 85 °C the conversion was 10 and 95%, respectively. At the same time, the content of ethane by-product was always <1%. Little activity was observed below 0.04% of Pd owing to the location of Pd in the interlayer spacing of the CNFs, making Pd unavailable for catalysis. An increase in Pd content up to 0.3 wt% resulted in a highly active catalyst. In bimetallic Pd–Ag catalysts, a beneficial effect of Ag was not observed. Apparently, Ag formed an alloy with Pd, which depleted the latter in the active phase.

In another study on the removal of acetylene from a mixture with ethylene, Chesnokov *et al.*[36] prepared several Pd catalysts supported on CNFs and CNTs. For CNFs, stacked and coaxial conical types were prepared. The experiments were conducted in a continuous-flow reactor using an ethylene–acetylene (99.5 : 0.5) mixture as the feed. The CNTs were prepared from a propane–butane mixture at 700 °C over an Fe–Mo–Co/Al_2O_3 catalyst. CNFs with a coaxial conical structure were prepared by methane decomposition on Ni/Al_2O_3 catalyst at 550 °C whereas CNFs with a stacked structure were prepared by methane decomposition on an Ni–Cu–Fe/Al_2O_3 catalyst at 700 °C. In addition, N-doped CNFs (3.5 wt% of N) (N-CNFs) were obtained by decomposition of a C_2H_4–NH_3 (40 : 60) mixture over Ni–Cu–Al_2O_3 catalyst at 550 °C. Before the catalyst preparation, the supports were treated with nitric acid, washed and dried. An aqueous solution of $PdCl_2 + HCl$ was used for the deposition of Pd on the supports. Figure 8.4 shows that a temperature of at least 100 °C was necessary to observe activity of the Pd/N-CNF catalyst yielding exclusively ethylene. The performance of this catalyst further improved with an increase in temperature above 100 °C and in Pd content from 0.05 to 0.15 wt%. Complete conversion to ethylene was observed for the catalyst containing 0.15 wt% of Pd. The Pd/CNF and Pd/CNT catalysts were more active below 100 °C but undesirable ethane was among the products. The superior performance of the Pd/N-CNF catalyst supported on coaxial conical CNFs was attributed to the modified electronic effect of the surface of the support caused by N doping. The superiority of this catalyst was confirmed by both high conversion and selectivity with respect to ethylene.

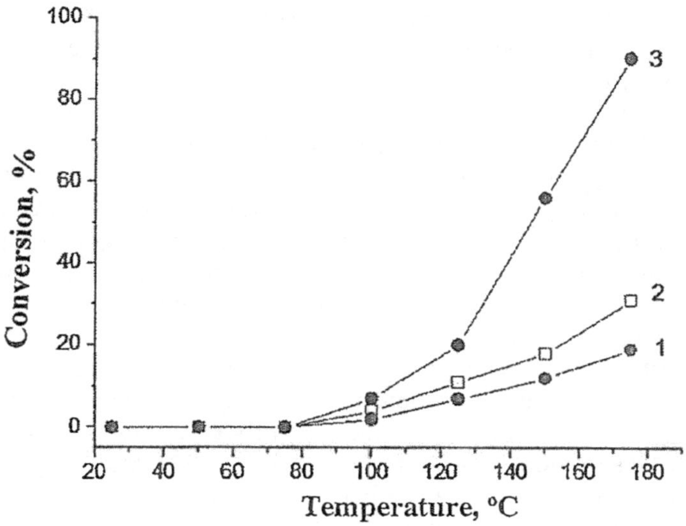

Figure 8.4 Effect of temperature and Pd content on acetylene conversion over Pd/CNF catalysts: 1, 0.05%; 2, 0.1%; 3, 0.15%.
Reproduced from ref. 36 with permission from Elsevier, Copyright 2017.

Table 8.3 Activity and selectivity of Pd catalysts in the HYD[a] of 1-octyne to 1-octene.[22]

Catalyst	Activity/mol (kg Pd)$^{-1}$ s^{-1}	Selectivity/%		
		1-Octene	1-Octane	Others
Pd/H-CNF	5.1	68.5	17.5	14.0
Pd/P-CNF	4.8	81.9	11.8	6.3
Lindlar[b]	5.7	88.6	7.4	4.0
Pd/AC	11.9	21.0	42.0	37.0

[a]Conditions: 294 K, 0.15 MPa H_2.
[b]Pd/$CaCO_3$.

Jung *et al.*[22] studied the HYD of 1-octyne to 1-octene rather than to octane over Pd catalysts supported on herringbone CNFs (H-CNFs) and platelet CNFs (P-CNFs). The catalysts were prepared by the colloidal microwave processing method. For comparison, Pd/$CaCO_3$ (Lindlar catalyst) and a commercial Pd/AC catalyst were used. The tests were conducted in a batch reactor at 0.15 MPa and 294 K. According to Table 8.3,[22] the Lindlar catalyst outperformed the Pd/CNF catalysts. The Pd/AC catalyst was the most active but its selectivity with respect to 1-octene was very low.

8.1.4 Chlorinated Hydrocarbons

Ordóñez *et al.*[37] compared Pd catalysts supported on untreated and nitric acid-treated CNFs in the HDCl of tetrachloroethylene. The catalysts were prepared by impregnation of the support using $PdCl_2$ precursor dissolved in either an aqueous or 2-propanol solution. The testing was conducted in a continuous fixed-bed reactor at 523 K and 0.6 MPa of H_2 using a solution of the reactant in toluene. The Pd/CNF catalyst prepared using the aqueous precursor exhibited higher initial activity but it deactivated at a greater rate than the Pd/CNF catalyst prepared when a 2-propanol solution of the precursor was used. For the latter catalyst, poisoning by chlorine was the main cause of catalyst deactivation whereas coke formation was the dominant cause for the Pd/CNF catalyst prepared using an aqueous solution of precursor. Also, more active catalysts were obtained using the nitric acid-treated CNFs compared with the untreated CNFs.

The effect of the type of carbon support on the performance of Pd catalysts in the HDCl of chlorobenzene was investigated by Amorim *et al.*[38] CNFs, ACs and graphite were used as carbon supports. The ACs and graphite were commercial products whereas the CNFs were prepared by the decomposition of ethylene over unsupported Ni. The CNFs were treated with nitric acid to remove Ni and oxidized in 5% O_2 + He at 673 K to remove amorphous carbon. The catalysts were prepared by impregnation of the supports using $Pd(NO_3)_2$ solution in isobutanol, then air dried at 393 K. Before the tests, the catalysts were reduced in a flow of H_2 at 523 K. The performances of the Pd/CNF, Pd/AC and Pd/graphite catalysts were determined under identical conditions in a continuous system with near atmospheric pressure H_2 at 423 K. Benzene was the main product of the HDCl of chlorobenzene with traces of cyclohexane formed during secondary HYD reactions. The following HDCl activity order of the catalysts was established: Pd/CNF > Pd/AC > Pd/graphite. In addition, Pd/CNF exhibited the highest stability, as indicated by a steady activity with time on-stream. Moreover, after regenerating the catalysts by treatment in H_2, the performance of the Pd/CNF catalyst during the subsequent cycle was the best.

Pd catalysts supported on graphite, CBs and CNFs were compared by Baeza *et al.*[39] in the aqueous-phase HDCl of 4-chlorophenol under mild conditions (*e.g.* 30 °C, atmospheric pressure H_2, batch system). Under these conditions, the highest HDCl activity was exhibited by the Pd/CB catalyst. The mean particle size of Pd on this catalyst approached 3 nm. Otherwise, among these catalysts the Pd particle size varied between 3 and 320 nm. Nevertheless, for these catalysts, the observed HDCl rates under such mild conditions deserve further attention.

8.2 Transition Metals Supported on CNFs

Similarly to the CNT-supported catalysts, the transition metals used for the preparation of conventional HPR catalysts (*e.g.* Ni, Co, Mo and W) were evaluated for the preparation of CNF-supported catalysts. Carbides, particularly Mo_2C and W_2C, supported on CNFs have been catalytic phases of interest in addition to metallic phases.

8.2.1 Conversion of α,β-Unsaturated Aldehydes

Catalysts consisting of Ni supported on herringbone CNFs (H-CNFs) and platelet CNFs (P-CNFs) and also on commercial ACs and one prepared by a spray-drying method (SD-AC) were compared in the HYD of CALD to HCALD followed by HYD (Figure 5.2) of the latter to HCALC.[40] Testing was performed in an autoclave at 160 °C and 3 MPa for 1 h. In this case, the rate constants for the scheme were compared. Table 8.4 confirms a high selectivity for the HYD of the C=C bond. This selectivity with respect to HCALD increased with decrease in the size of the Ni particles regardless of the type of support used. The Ni/P-CNF catalyst was the most selective for the HYD of the C=C bond and for the overall conversion.

Gryglewicz *et al.*[41] compared herringbone CNFs with N-doped CNFs as supports for Ni catalysts used for the conversion of citral in toluene. Testing was performed in a batch reactor at 150–200 °C and 3 MPa (Figure 5.3). The CNFs were prepared by the CVD method at 550 °C using methane as carbon source over Ni/Al_2O_3 catalyst. The N doping of the CNFs was carried out by ammonia treatment of nitric acid-oxidized CNFs. The Ni catalysts were prepared by incipient wetness impregnation of CNFs using an aqueous solution of Ni nitrate. For both catalysts, more than 96% selectivity with respect to citronellal was observed at 150 °C, whereas at 200 °C a secondary reaction involving the cyclization of citronellal to isopulegol became evident. Over the N-doped CNF-supported Ni catalyst, the HYD of *cis*-citral relative to *trans*-citral was favored. Over this catalyst, the overall HYD rate was 1.5 times higher than that over the undoped catalyst.

The HYD of crotonaldehyde to crotyl alcohol was used as a model reaction to study the activity of Ni(5%) supported on CNFs and Al_2O_3 in a continuous reactor in a flow of H_2 at 75–150 °C.[42] Two types of CNFs with different orientations, *i.e.* platelets and ribbons (Figure 2.4), were used. The Ni catalysts were prepared by incipient wetness impregnation using an ethanolic solution of the Ni precursor. The effect of temperature on catalyst activity is shown in Figure 8.5. The performance of the Ni catalysts supported on CNFs with a platelet orientation was much better than that on CNFs with a ribbon orientation. The selectivity with respect to crotyl alcohol increased with increase in temperature from 75 to 150 °C, *i.e.* for Ni/platelet and Ni/ribbon

Table 8.4 Rate constants for the HYD[a] of CALD over Ni supported on CNFs and ACs.[40]

Catalyst	Rate constant/min^{-1}		
	$k_{C=C}$	$k_{C=O}$	$k_{C=C}/k_{C=O}$
Ni/H-CNF	0.0261	0.0008	32.6
Ni/P-CNF	0.0383	0.0015	25.5
Ni/AC	0.0174	0.0004	43.5
Ni/SD-AC	0.0341	0.0019	17.9

[a]Conditions: 160 °C, 3 MPa H_2.

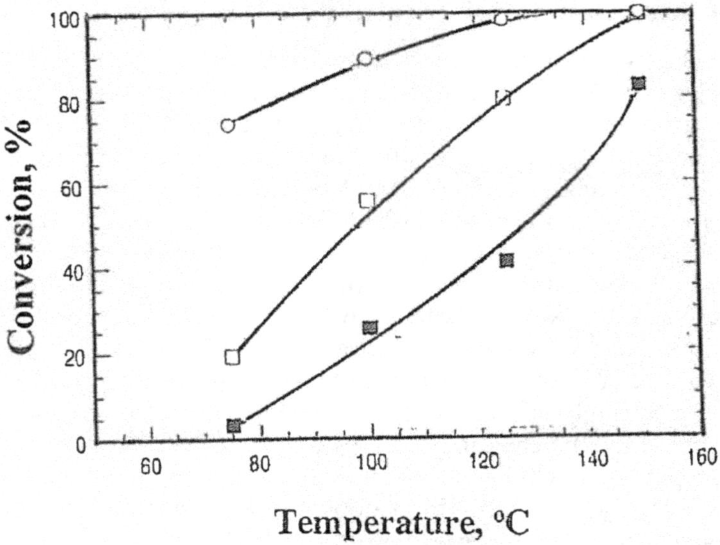

Figure 8.5 Effect of CNF type and temperature on the conversion of crotonaldehyde: ○, Ni/ribbon CNF; □, Ni/platelet CNF; ■, Ni/Al_2O_3.
Reproduced from ref. 42 with permission from Elsevier, Copyright 1999.

catalysts, the selectivity increased from ~13 to ~78% and from ~1 to ~74%, respectively. Both the activity and selectivity of the Ni/Al_2O_3 catalyst were much lower.

8.2.2 Conversion of Lignocellulosic Reactants

Van de Vyver *et al.*[43] used an Ni/CNF catalyst with Ni located at the tips of the CNFs for the conversion of cellulose to sugar alcohols. The catalyst was prepared by the CVD of methane over an Ni/Al_2O_3 catalyst. For comparison, Ni/AC and Ni/Al_2O_3 catalysts were also used. The tests were conducted in a batch system at 463–503 K and 2–6 MPa of H_2. Over the Ni/CNF catalyst, the selectivity with respect to sorbitol approached 50% at an overall cellulose conversion of 92%. In this regard, the performance of the Ni/AC and Ni/Al_2O_3 catalysts was much lower. The high activity and selectivity of the Ni/CNF catalysts were attributed to the easy access of cellulose molecules to Ni located at the edge regions of the CNFs. In another study by Van de Vyver *et al.*,[44] CNFs prepared by the CVD of methane on an Ni/Al_2O_3 catalyst were treated with nitric acid to remove Ni. After washing and drying, Ni was deposited on the CNFs by incipient wetness impregnation with an aqueous solution of $Ni(NO_3)_2$. After drying at 333 K and heat treatment at 700 K in N_2, the Ni/CNF catalyst was reduced in a flow of H_2. The experiments were carried out in a batch reactor at 463 K and 6 MPa of H_2. Under these conditions, almost 70% selectivity with respect to sorbitol at 92% conversion of cellulose was obtained.

Jongerius *et al.*[45] compared the performances of W_2C and Mo_2C catalysts supported on CNFs. The catalyst activity and selectivity were evaluated in the HDO of guaiacol in a batch reactor at 300–375 °C and 5.5 MPa. The W_2C/CNF catalyst was prepared by the impregnation of CNFs using ammonium metatungstate solution to obtain 15 wt% W. After impregnation, the sample was vacuum treated at 80 °C for 24 h, followed by heat treatment at 1000 °C under N_2 for 3 h. The Mo_2C/CNF catalyst was prepared by impregnation using ammonium molybdate solution to obtain 7.5 wt% Mo, followed by heat treatment at 900 °C under N_2. The conversions and product distributions are given in Table 8.5.[45] For both Mo_2C/CNF and W_2C/CNF catalysts, high selectivity with respect to phenolics was observed. This is in line with the high selectivity of unsupported Mo_2C and W_2C for the hydrogenolysis of C–N, C–S and C–O bonds (without HYD of aromatic rings) observed for various heteroring-containing reactants.[2] This differs from the conventional Co(Ni)/Mo(W)/Al_2O_3 catalysts, for which the hydrogenolysis of the heterobonds requires HYD of the aromatic ring as the first step. Therefore, the metal carbides appear to be the catalysts of choice when the depolymerization of biomass to monomeric phenols is the objective.

Santillan-Jimenez *et al.*[46] used CNFs, CNTs and ACs as supports for the preparation of Mo_2C catalysts. Ammonium molybdate (Am) and molybdic acid (Ac) were used as Mo precursors to obtain either 7.5 or 20 wt% of Mo. The catalysts were used in the HDO of guaiacol in dodecane. The testing was conducted in an autoclave at 300 °C and ~4 MPa of H_2 for 4 h. The objective was to maximize the yield of phenol. In this regard, the best performance was exhibited by the 20% Ac/CNF, 7.5% Am/CNF, 20% Am/CNF and 20% Am/CNT catalysts (Table 8.6). Under identical conditions over a reference catalyst such as Ru(5%)/AC, quantitative conversion of guaiacol with 100% selectivity with respect to cyclohexane was observed.

Moreira *et al.*[47] studied the conversion of guaiacol over Mo_2C/CNF catalysts. For the preparation of the catalysts, commercial CNFs (as received) were treated in boiling nitric acid. After filtration, the CNFs were washed with distilled water and dried. The catalysts were prepared by incipient wetness impregnation with ammonium heptamolybdate solution, followed by carburization at 1023 K in a flow of H_2. The tests were conducted in a batch reactor using a solution of guaiacol in decane at 573 and 623 K and

Table 8.5 Conversion and product distribution in the HDO of guaiacol.[45]

Catalyst	Conversion/%	Product distribution[a]/%				
		A	B	C	D	E
None	40	7	8	<1	2	4
CNF	40	7	11	<1	3	4
W_2C/CNF	66	46	12	<1	4	2
Mo_2C/CNF	>99	45	13	2	9	2

[a]A, phenol; B, *o*-cresol, *p*-cresol and dimethylphenol isomers; C, benzene and toluene; D, anisole, methylated anisole and dimethoxybenzene; E, cyclohexane, cyclohexene and cyclohexanone.

Table 8.6 Conversion and selectivity of Mo_2C catalysts in upgrading of guaiacol in dodecane over different catalysts.[46]

Catalyst[a]	Conversion/%	Selectivity/% Catechol	Phenol	Mass balance[b]
None	7	15	0	98
AC	12	10	3	99
Ac(7.5%)/AC	26	21	12	98
Ac(20%)/AC	61	7	9	70
Am(7.5%)/AC	37	20	1	98
Am(20%)/AC	53	31	3	95
CNF	6	19	3	97
Ac(7.5%)/CNF	40	8	12	85
Ac(20%)/CNF	56	9	35	96
Am(7.5%)/CNF	33	9	38	98
Am(20%)/CNF	53	11	37	96
CNT	16	11	2	97
Ac(7.5%)/CNT	72	36	3	95
Ac(20%)/CNT	91	24	5	98
Am(7.5%)/CNT	33	33	3	97
Am(20%)/CNT	24	9	36	96

[a]Conditions: 0.125 g catalyst, 1.25 g guaiacol, 11.25 g dodecane, 300 °C, 4 MPa H_2, 4 h.
[b]Mass balance includes only liquids and solids at the start and end of the reaction.

2 MPa of H_2. At 523 K, phenols were the predominant products because of an insignificant demethoxylation reaction. Therefore, that the demethylation was of guaiacol was quite evident. Whereas with CNFs alone the conversion reached about 5% at 573 K after 2 h, over the Mo_2C/CNF catalyst the conversion more than doubled, giving phenol and xylenols as the major products. At 623 K, after 2 and 4 h the conversion over the catalyst reached 50 and 80%, respectively.

Ochoa *et al.*[48] evaluated an Mo_2C catalyst supported on CNFs in the HDO of guaiacol dissolved in decane at 300 °C and 3.2 MPa of H_2 in an autoclave The CNFs were prepared by the decomposition of a 50 : 50 mixture of CH_4 and CO_2 over an Ni–Co–Al catalyst. For catalyst preparation, the purified CNFs were impregnated with an aqueous solution of ammonium heptamolybdate. This solid was carburized in a flow of H_2 at 550, 600, 650, 700 and 750 °C. Both the amount of Mo_2C/CNF and its crystallite size were influenced by the carburization temperature. With respect to the Mo_2C crystal size, the heating rate was the main parameter. Large Mo_2C crystallites were obtained at slow heating rates. The most active Mo_2C/CNF catalyst was prepared at a heating rate of 1 °C min^{-1} and a carburization temperature of 750 °C.

Remón *et al.*[49] prepared Ni–Co/CNF catalysts and applied them for the upgrading of bio-oil derived from lignocellulosic biomass. The tests were conducted in a batch reactor under supercritical water conditions (*e.g.* 380 °C, 23 MPa). Prior to catalyst preparation, the CNFs were oxidized *via* nitric acid pretreatment, followed by thermal treatment before the incorporation of the metal active phase. Optimal quality and yield of liquid products could be attained at a temperature of thermal treatment that yielded a moderate amount of O-containing groups on the surface of the CNFs. The stability of the CNF support under the supercritical water conditions was not investigated.

8.2.3 Fatty Acids and Esters

Hollak *et al.*[50] prepared CNFs by flowing a CO–H_2–N_2 mixture over an Ni/SiO_2 catalyst at 550 °C for 24 h. The CNFs were treated in boiling 1 M KOH to remove SiO_2. After thorough washing, the CNFs were treated with nitric acid to remove Ni. The support was impregnated with ammonium metatungstate to obtain 15 wt% of W. In parallel, the CNFs were impregnated with ammonium molybdate to obtain a 7.5 wt% loading of Mo. This ensured similar atomic amounts of the metals. Before testing, the W and Mo catalysts were heat treated in N_2 at 1000 and 900 °C, respectively. The catalysts were applied in the HDO of oleic acid in dodecane in a batch reactor at 350 °C and 5 MPa of H_2. Under these conditions, the Mo_2C/CNF catalyst exhibited a higher HYD activity, as indicated by a lower content of olefinic products than that obtained over the W_2C/CNF catalyst. The deactivation of the latter by coke deposition was more pronounced.

The Mo_2C/CNF catalysts used in the HDO of methyl palmitate by Qin *et al.*[51] were prepared by carbothermal H_2 reduction of ammonium molybdate-impregnated CNFs. This included carburization in H_2 in three steps from 450 to 700 °C. The experiments were conducted in an autoclave at 240 °C and 2 MPa of H_2 using the reactants dissolved in hexane. The Mo_2C/AC and Mo_2C/CNT catalysts used for comparison were prepared using similar procedures. The effect of the amount of Mo_2C on the conversion and selectivity, defined as the yield of hydrocarbons, is shown in Figure 8.6. Optimal catalyst performance was attained with 20 wt% of Mo_2C. The effect of the support on the activity and selectivity of the catalysts containing 20 wt% of Mo_2C is shown in Table 8.7. Among these catalysts, the differences in conversion and selectivity with respect to hydrocarbons were minor. However, a high yield of hexadecene obtained over the Mo_2C/CNT catalyst indicated a lower HYD activity compared with the other catalysts.

Gosselink *et al.*[52] prepared a series of W_2C(15%)/CNF catalysts by the impregnation of CNFs using an aqueous solution of ammonium metatungstate. After separation, the solid was dried at 80 °C in vacuum before being treated under N_2 in the temperature range 600–1000 °C. The evolution of the W_2C phase with increase in temperature was evident. At the same time, the amount of WO_3 decreased. The catalysts were tested in the conversion of stearic acid dissolved in dodecane (350 °C, 5 MPa, batch process). In this case, n-C_{17} and n-C_{18} hydrocarbons were the main products. The catalysts containing WO_3 were selective for n-C_{17} formation and those containing W_2C for n-C_{18} formation.

Mo_2C/CNF and W_2C/CNF catalysts prepared by impregnation of nitric acid-oxidized CNFs and hydrochloric acid-treated CNFs were compared in the HDO of stearic acid to n-C_{17} and n-C_{18} hydrocarbons.[53] The C_{17}

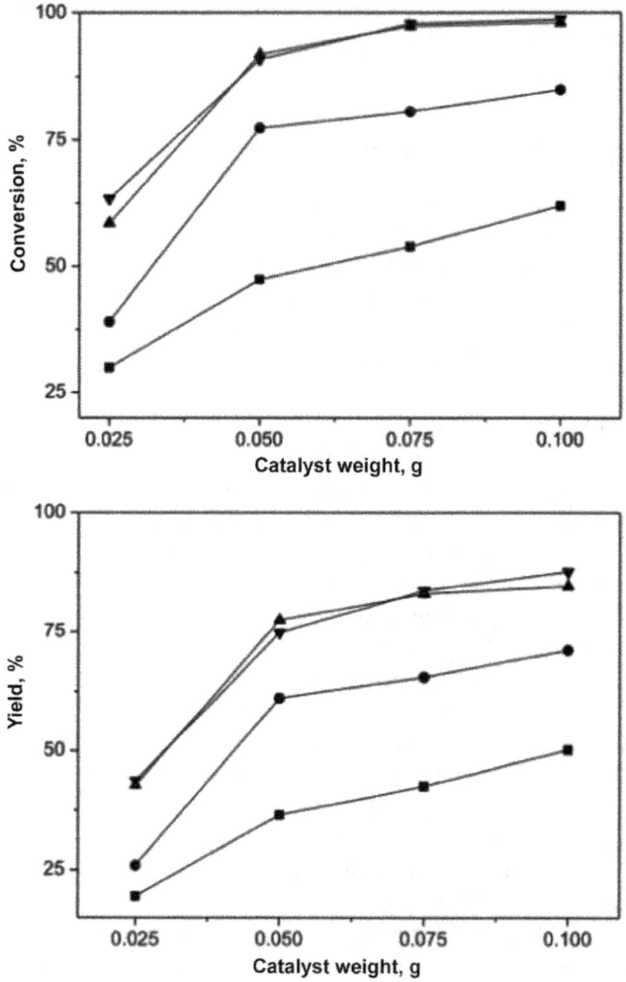

Figure 8.6 Effect of Mo_2C content on (top) the conversion of methyl palmitate and (bottom) the yield of hydrocarbons: ■, 5; ●, 10; ▲, 20; ▼, 40 wt%. Reproduced from ref. 51 with permission from the Royal Society of Chemistry.

Table 8.7 Activity and selectivity of Mo_2C (20%) catalysts in the HDO of methyl palmitate.[51]

			Selectivity[b]/%			
Catalyst	Conversion/%	Yield[a]/%	A	B	C	D
Mo_2C/CNF	58.6	42.8	8.3	1.2	33.2	0
Mo_2C/AC	55.7	41.1	0.9	33.4	6.3	0.4
Mo_2C/CNT	56.2	42.6	0.8	25.7	14.8	1.1

[a]Hydrocarbons.
[b]A, *n*-heptadecane; B, *n*-hexadecane; C, hexadecene; D, isoalkanes.

hydrocarbons were formed from C_{18} oxygenated intermediates by dehydrogenation–decarbonylation reactions. For the experiments conducted in a batch reactor (200–350 °C, 3 MPa of H_2), the reactant was dissolved in dodecane. The catalyst performance increased with increasing size of the metal carbide particles, *e.g.* on increasing the size from 2 to 10 nm, the activity of the catalysts almost doubled. The resistance of the catalysts to deactivation also improved with increasing particle size. This is in an agreement with the performance of unsupported, bulk metal carbides.[54] In terms of conversion, the W_2C catalyst outperformed the M_2C catalyst.

Molybdenum carbide (α-MoC_{1-x}) supported on CNFs was prepared by carburization using an H_2–CH_4 mixture and β-Mo_2C/CNF by carbothermal reduction in argon.[55] Prior to these steps, the CNF support was impregnated with ammonium molybdate solution by incipient wetness impregnation. The activities and selectivities of the catalysts were compared in the conversion of stearic acid to hydrocarbons. The α-MoC_{1-x}/CNF catalyst exhibited a better performance, as indicated by the 80% conversion reached after 4 h. With the β-Mo_2C/CNF catalyst, a similar conversion was reached after 6 h. The activity of the CNF support was an order of magnitude lower than that of the catalysts. The experiments were carried out in a batch reactor at 350 °C and 4 MPa of H_2.

8.2.4 Conversion of Hydrocarbons

Cu–Fe bimetallic catalysts supported on CNFs, ACs and γ-Al_2O_3 were compared in the HYD of ethylene at 80 °C and atmospheric pressure H_2 in a continuous system.[56] Under these conditions, the Cu–Fe/CNF catalysts outperformed the other catalysts. In other work, 1-butene and 1,3-butadiene were studied under similar conditions using Ni catalysts supported on the same supports.[57,58] Again, the Ni/CNF catalyst exhibited much higher activity than the Ni/AC and Ni/Al_2O_3 catalysts. The best performance of the Ni/CNF catalyst could not be reconciled with the particle size distribution on the supports. Apparently, the high activity of the Ni/CNF catalyst could be attributed to modifications of the Ni particle characteristics facilitated by the chemical and structural properties of the CNF support. For AC and γ-Al_2O_3 supports, such effects on Ni particles were not evident. Three types of CNFs (Figure 2.4), *i.e.* platelet, ribbon and spiral, were prepared with the aim of further optimizing the performance of Ni/CNF catalysts.

CNFs prepared by Liu *et al.*[59] *via* the catalytic decomposition of ethylene had a mean diameter of 50–60 nm and lengths up to several tens of micrometers. These CNFs were used as supports for the preparation of an Ni catalyst, using incipient wetness impregnation of CNFs with a solution of Ni precursor. The size of the Ni particles in the resultant Ni/CNF catalyst varied from 10 to 20 nm. The catalyst exhibited good activity for the HYD of benzene (180 °C, batch reactor) with almost 100% selectivity with respect to cyclohexane.

Fishbone CNFs functionalized with nitric acid were used as the support for Ni–Mo catalysts in the HYD of anthracene in a batch reactor at 300 °C and 7 MPa H_2.[60] The severity of the oxidation treatment influenced the size of the MoS_2 slabs formed on sulfidation. Milder treatment resulted in short and defective slabs, ensuring high HYD activity. On the other hand, severe treatment yielded long stacked slabs exhibiting lower HYD activity.

8.2.5 Petroleum-derived Reactants and Feeds

Comprehensive evaluations of catalysts containing conventional metals such as Co, Ni and Mo supported on CNFs were conducted by Yu *et al.*[61] As catalyst supports, fishbone and platelet CNFs (Figure 2.4) were used in addition to γ-Al_2O_3. The fishbone CNFs were prepared by CO decomposition on an Ni/SiO_2 catalyst and the platelet CNFs by CO decomposition on an Fe_3O_4 catalyst. Both CNFs were functionalized and purified in boiling concentrated nitric acid. In addition, the fishbone CNFs were treated with a solution of NaOH to remove SiO_2. After washing and drying, the supports were used for the preparation of monometallic Co, Ni and Mo catalysts and bimetallic Co–Mo and Ni–Mo catalysts. In this, case, a standard incipient wetness impregnation method was used. The catalysts were applied in the HDS of thiophene in a continuous fixed-bed microreactor. Before the experiments, the catalysts were sulfided in a flow of H_2S(5%)–H_2 at 407 °C. The results are summarized in Table 8.8. For both CNFs, the monometallic Co catalysts exhibited higher activity than the bimetallic Co–Mo and Ni–Mo catalysts. However, among all the catalysts tested, the sulfided Ni–Mo/Al_2O_3 catalyst was the most active. Similarly, in a study by Prajapati and Verma,[62] the following activity order in the HDS of thiophene was established: Ni–Mo/Al_2O_3 > Ni–Mo/platelet-CNF > Ni–Mo/fishbone-CNF. It should be noted that sulfiding of the metals modified the

Table 8.8 Activity of catalysts in the HDS of thiophene.[61]

Catalyst composition/%			
Mo	Co	Ni	Conversion[a]/%
Fishbone CNFs			
12	—	—	3.8
—	12	—	19.2
12	4	—	12.7
12	—	4	17.8
Platelet CNFs			
—	12	—	23.0
12	4	—	11.9
12	—	4	18.1
γ-Al_2O_3			
12	—	—	5.5
12	4	—	12.6
12	—	4	26.5

[a]Conversion after 6.5 h on-stream.

interaction with the supports. Presumably, the beneficial effects of the supports on the activity of metals observed in many studies were minimized by sulfiding of the catalysts as performed in the present study.[62]

Pinilla *et al.*[63] prepared nanosheets of MoS_2 decorated with Ni by decomposition of ammonium thiomolybdate on functionalized CNFs (FCNFs), followed by impregnation with nickel nitrate. The decomposition temperature influenced the morphology of MoS_2 crystallites, *e.g.* more homogeneous MoS_2 slabs were obtained at 450 than at 600 °C. Consequently, the former catalyst was more active. Ni–Mo/CNF catalysts were compared with conventional Ni–Mo/Al_2O_3 catalysts using a vacuum residue (VR) derived from petroleum. The tests were performed in a batch reactor. A mixture of VR (0.5 g), catalyst (0.12 g) and CS_2 (0.1 mL) was treated at 425 °C and 18.5 MPa of H_2 for 1 h. The conversions of the residue (>450 °C to <450 °C) and that of asphaltenes and other conversion parameters are given in Table 8.9.[63] These results showed that the Ni–Mo/FCNF-450 catalyst outperformed the conventional Ni–Mo/Al_2O_3 catalyst only during the conversion of asphaltenes. The latter catalyst had more suitable textural properties for HDM (hydrodemetallization) in spite of the much greater amount of coke deposited on the catalyst surface. This suggests that the CNF-supported catalysts were more efficient in activating hydrogen required for the stabilization of coke precursors. However, runs of longer duration in a continuous reactor would be necessary to confirm the differences in activity. Moreover, optimization of the catalyst preparation and pretreatment procedures may change the relative activities.

AL-Ahmadi *et al.*[64] used γ-Al_2O_3 alone and a γ-Al_2O_3–CNF composite as supports for the preparation of Co(5%)–Mo(15%) catalysts. The composite was prepared by mixing γ-Al_2O_3 and CNF using a sol-gel method. Subsequently, the mixture was refluxed at 110 °C in deionized water, ethanol and diethylene glycol medium. After separation, the solid obtained was dried at 100 °C. The Co–Mo catalysts were prepared by incipient wetness impregnation. The activity of the catalysts was evaluated in the HDS of DBT in a batch reactor at 300 °C and 5.5 MPa of H_2. DBT was dissolved in decalin, giving 550 ppm of sulfur. With respect to the HDS rate, the Co–Mo/Al_2O_3–CNF catalysts exhibited much higher activity than the Co–Mo/Al_2O_3 catalyst. The catalyst characterization revealed that the former had more suitable textural properties such as surface area and porosity. Moreover, TEM confirmed that in the Co–Mo/Al_2O_3–CNF catalyst, the active metals were associated with CNFs rather than with Al_2O_3. This ensured a more efficient dispersion of active metals in addition to beneficial electronic effects conferred by CNFs on active metals.

Table 8.9 Conversion (>450 °C to <450 °C) parameters.[63]

Sample	$C_{>450}$	C_{asph}	HDS	HDM	C_{Ni}	C_V	Coke/%
Ni–Mo/FCNF-450	0.51	0.73	0.83	0.86	0.67	0.94	4.7
Ni–Mo/FCNF-600	0.49	0.57	0.78	0.74	0.46	0.85	8.1
Ni–Mo/Al_2O_3	0.55	0.59	0.9	0.97	0.95	0.98	28.4

8.3 Other Metals Supported on CNFs

ReO_x/CNF catalysts with the ReO_x loading varying from 2 to 13 wt% were applied in the HDO of phenol in a batch reactor at 573 K and H_2 pressure from near atmospheric up to 5 MPa.[65] The optimal ReO_x coverage, confirmed by the highest activity, was attained at 10 wt%. This corresponded to the reduction of ReO_x to Re^{4+} and Re.[6,7] The catalyst exhibited high selectivity for C=O bond cleavage leading to the formation of benzene. Similarly, high selectivity with respect to aromatic hydrocarbons (benzene, toluene and xylenes) was observed in the HDO of guaiacol, anisole and *o*-cresol. When the H_2 pressure approached 5 MPa, the formation of hydrogenated products became evident.

8.4 Summary of Studies on CNF-Supported Catalysts

The combined database on the development of CNF-supported catalysts in Table 8.10 is much less extensive than that on catalysts supported on CNTs (Tables 7.8, 7.22, 7.24, 7.31 and 7.33). Both noble metal- and conventional metal-containing catalysts supported on CNFs are summarized in Table 8.10. Except for two studies conducted in continuous fixed-bed reactors, all other studies were carried out in batch systems. Overall, similar reactants and feeds have been tested. However, there are a few studies in which catalysts supported on CNTs were compared with those supported on CNFs under the same conditions. Ledoux *et al.*[25] reported that in the HYD of CALD in a batch reactor (353 K, 1 atm H_2), Pd/CNT exhibited a better performance than the Pd/CNF counterpart. However, this was observed under only one set of experimental conditions. Even these results (Figure 8.2) do not provide clear evidence for a better performance of one support over the other. Therefore, optimization during catalyst preparation and testing would be necessary to confirm the advantage of CNT supports compared with CNF supports under a wider range of experimental conditions. For example, the conditions of catalyst preparation have significant effects on activity and selectivity. Thus, in the preparation of Mo_2C catalysts supported on CNFs, CNTs and ACs, a molybdic acid precursor used for support impregnation gave more active catalysts than an ammonium molybdate precursor for all three supports.[43] In the HYD of guaiacol, the best performance was exhibited by an Mo_2C/CNT catalyst. On the other hand, in the HYD of CALD to CALC (Figure 5.2), a Pt/CNF catalyst outperformed a Pt/CNT catalyst.[22] In addition, over Mo_2C/CNF, Mo_2C/CNT and Mo_2C/AC catalysts, little difference in catalyst performance in the HYD of palmitic acid was observed.[50] In another example, in the selective HYD of acetylene to ethylene, a superior performance of a Pd catalyst supported on coaxial conical CNFs compared with one supported on CNFs with stacked graphene layers (Figure 2.4) was noted.[36] The former Pd/CNF catalyst was more active than the Pd/CNT counterpart. Therefore, a general order of catalyst performance based on activity and selectivity cannot be established without extensive catalyst evaluations.

Table 8.10 Summary of studies on catalysts supported on CNFs.

Catalyst	Reactant	Conditions	Ref.
Noble metal-containing catalysts			
Oxygenates			
1. Pt/CNF; Ru/CNF	CALD	345 K; 1 atm; 2-propanol–water	19
2. Pd/CNF; Pd/AC	CALD	353 K; 1 atm; dioxane	20, 21
3. Pt/CNF; Pt/CNT; Pt/AC; Pt/Al_2O_3	CALD	383 K; 5 MPa; batch; THF	22
4. Pt/CNF; Ru/CNF	CALD	383 K; 4.8 MPa; 2-propanol	23, 24
5. Pd/CNF; Pd/CNT; Pd/AC	CALD	353 K; 1 atm; dioxane	25
6. Pd/CNF–TiO_2	Citral	363 K; 2 MPa; 2-propanol	26–28
7. Pt/CNF; Pt/AC	Citral	353–373 K; 0.5 and 1 MPa; hexane	29
8. Ru/CNF–graphite	Sorbitol	493 K; 8 MPa; aq.	30
9. Ru/CNF	Sorbitol	493 K; 8 MPa; aq.	31
10. Ru/CNF; Ru/AC	Sorbitol	493 K; 2 MPa; aq.	32
11. Ru/CNF; Ru/AC	LA	432 K; 4.5 MPa; aq.	33
Hydrocarbons			
12. Pt/CNF	*n*-Hexane	523 K; 1 atm; continuous	34
13. Pd/CNF; Pd–Ag/CNF	Acetylene	308–423 K, 1 atm; continuous	35
14. Pd/CNF; Pd/CNT	Acetylene	333–453 K; 1 atm; continuous	36
15. Pd/CNF; Pd/AC	1-Octyne	294 K; 0.2 MPa; batch; heptane	22
Chlorohydrocarbons			
16. Pd/CNF	Chloroethene	523 K; 0.6 MPa; continuous; toluene	37
17. Pd/CNF; Pd/AC; Pd/graphite	Chlorobenzene	423 K, 1 atm; continuous	38
18. Pd/CNF; Pd/AC; Pd/graphite	Chlorophenol	303 K; 1 atm; aq.	39
Conventional metal-containing catalysts			
Oxygenates			
19. Ni/CNF; Ni/AC	CALD	433 K; 3 MPa; batch; toluene	40
20. Ni/CNF	Citral	423–473 K; 3 MPa; batch; toluene	41
21. Ni/CNF; Ni/Al_2O_3	Crotonaldehyde	348–423 K; 1 atm; continuous	42
22. Ni/CNF; Ni/AC; Ni/Al_2O_3	Cellulose	463–503 K; 2–6 MPa; batch; aq.	43, 44
23. W_2C/CNF; Mo_2C/CNF	Guaiacol	573–648 K; 5.5 MPa; batch; dodecane	45
24. Mo_2C/CNF; Mo_2C/CNT; Mo_2C/AC; Ru/AC	Guaiacol	573 K; 4 MPa; batch; dodecane	46

Table 8.10 (*Continued*)

Catalyst	Reactant	Conditions	Ref.
25. Mo_2C/CNF	Guaiacol	573 and 623 K; 2 MPa; decane	47
26. Mo_2C/CNF	Guaiacol	573 K; 3.2 MPa; batch; decane	48
27. Ni–Co/CNF	Bio-oil	In supercritical water	49
28. Mo_2C/CNF; W_2C/CNT	Oleic acid	623 K; 5 MPa; batch; dodecane	50
29. Mo_2C/CNF; Mo_2C/CNT; Mo_2C/AC	Methyl palmitate	513 K; 2 MPa; batch; hexane	51
30. W_2C(15%)/CNF	Stearic acid	623 K; 5 MPa; batch; dodecane	52
31. Mo_2C/CNF; W_2C/CNF	Stearic acid	473–623 K; 3 MPa; batch; dodecane	53
32. Mo_2C/CNF	Stearic acid	623 K; 4MPa H_2; batch; dodecane	55
Other reactants and feeds			
33. Cu–Fe/CNF; Cu–Fe/AC; Cu–Fe/Al_2O_3; Ni/CNF	Ethylene, butene, butadiene	353 K; 1 atm; continuous	56
34. Ni/CNF	Benzene	453 K; batch	59
35. Ni–Mo/CNF	Anthracene	573 K; 7 MPa; batch	60
36. Ni–Mo/CNF; Co–Mo/CNF; Ni–Mo/Al_2O_3	Thiophene	573 K; 1 atm; continuous	61, 62
37. Ni–Mo/CNF; Ni–Mo/Al_2O_3	Asphaltenes	698 K; 18.5 MPa; batch	63
38. Co–Mo/Al_2O_3–CNF; Co–Mo/Al_2O_3	DBT	573 K; 5.5 MPa; decalin; batch	64
39. ReO_x/CNF	Phenol, anisole, guaiacol, *o*-cresol	573 K; 0.1–5 MPa; batch	65

Potentially, the greater interest in CNT supports than in CNF supports may be attributed to the difference in cost and/or ease of preparation. This suggests that the former may be prepared more readily than CNF supports. However, this would have to be confirmed by some additional investigations. For CNT-supported catalysts, much more interest has been shown in noble metal-containing catalysts than in conventional metal-containing catalysts. In contrast, the amount of interest in noble metals and conventional metals supported on CNFs was similar.

Among the oxygenates in Table 8.10, some reactants were tested over both noble metals and conventional metals supported on CNFs. In every case, the severity of the experimental conditions over the latter catalysts was higher. This confirmed that noble metals are more active during HYD, presumably because of more efficient hydrogen activation compared with that over transition metals.[66]

In the case of CALD and citral, the mild conditions over noble metal-containing catalysts can be attributed to the partial conversion only of these reactants. For example, CALD was converted either to CALC or HCALD rather than to hydrocarbons, such as occurs during complete conversion. Generally, when hydrocarbons are the targeted products, the conversion of oxygenates requires more severe conditions, *i.e.* higher temperature and H_2 pressure and higher hydrogen consumption. In this regard, the same observations can be made also for other reactants in Table 8.10.

It has been observed that in every study in which catalysts supported on CNFs were compared with those supported on ACs and CBs and on oxidic supports, the CNF-supported catalysts exhibited better activity and selectivity. Similar observations have been made for CNT-supported catalysts. Moreover, it is believed that for both CNT- and CNF-supported catalysts, there is a high potential for further optimization of catalyst performance.

8.5 Catalysts Supported on CNHs

As indicated earlier, CNHs have attracted much less attention than CNTs and CNFs as supports for catalysts used in HYD and HPR applications. Similarly to the other carbon nanosupports, functionalization and/or pre-oxidation are desirable pretreatments to make CNHs suitable catalyst supports. As pointed out earlier, the opening of both ends of CNHs results in improvements in porosity and surface area. Thus, both the interior and exterior of CNHs become available for catalysis when both ends can be opened (Figure 2.5).

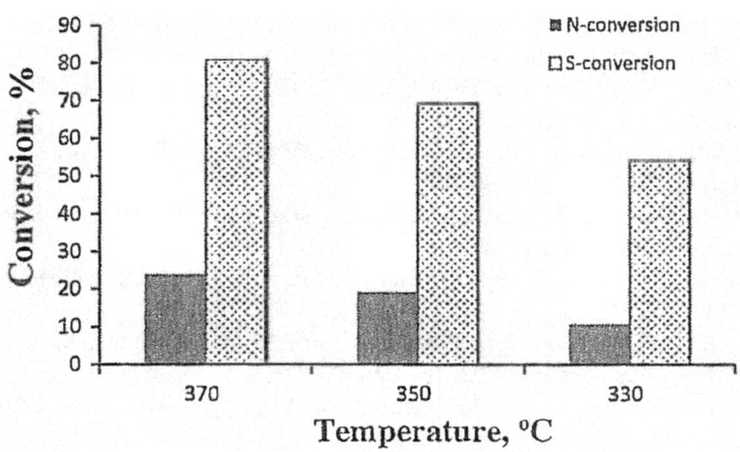

Figure 8.7 Effect of temperature on (○) HDS and (■) HDN of gas oil over an Ni–Mo/CNH catalyst.
Reproduced from ref. 67 with permission from Springer Nature, Copyright 2013.

Compared with catalysts supported on CNTs and CNFs, those supported on CNHs have received much less attention. So far, only one study relevant to HYD could be found in the literature. In that study, Aryee *et al.*[67] used nitric acid-functionalized CNHs as supports for a series of Ni(2.5%)–Mo(13%) catalysts. The duration of the acid treatment influenced the properties of the CNHs. At the optimal duration, the best performance of catalysts prepared by a co-impregnation method could be achieved. After presulfiding, the catalyst was precoked using a coker light gas oil before being used in the HPR of Athbasca gas oil in a continuous system at 330–370 °C and 8.8 MPa. Figure 8.7[67] shows an increase in HDS and HDN conversion with increase in temperature. It is evident that compared with HDN, the catalyst selectivity for HDS was much greater.

References

1. E. Bailón-García, F. J. Maldonado-Hódar, A. F. Pérez-Cadenas and F. Carrasco-Marín, *Catalysts*, 2013, **3**, 853–877.
2. C. Pham-Huu and M.-J. Ledoux, *Top. Catal.*, 2006, **40**, 49–55.
3. W. Deng, X. Tan, W. Fang, Q. Zhang and Y. Wang, *Catal. Lett.*, 2009, **133**, 167–174.
4. W. Deng, Y. Wang, Q. Zhang and Y. Wang, *Catal. Surv. Asia*, 2012, **16**, 91–105.
5. M. L. Toebes, J. M. P. van Heeswijk, J. H. Bitter, A. J. van Dillen and K. P. de Jong, *Carbon*, 2004, **42**, 307–315.
6. L. S. Ribeiro, J. J. Delgado, J. J. de Melo Órfão and M. F. Ribeiro Pereira, *ChemCatChem*, 2017, **9**, 888–896.
7. B. F. Machado, H. T. Gomes, P. Serp, P. Kalck and J. L. Faria, *ChemCatChem*, 2010, **2**, 190–197.
8. N. Oger, Y. F. Lin, C. Labrugère, E. Le Grognec, F. Rataboul and F.-X. Felpin, *Carbon*, 2016, **96**, 342–350.
9. J. Cheng, Y. Qiu, J. Zhang, R. Huang, W. Yang and Z. Fan, *Bioresour. Technol.*, 2017, **244**, 569–574.
10. A. P. Koskina, Y. V. Laricheva, A. I. Lysikova, O. N. Primachenko and S. S. Ivancheva, *Kinet. Catal.*, 2017, **58**, 655–662.
11. R. Liu, J. Chen, X. Huang, L. Chen, L. Ma and X. Li, *Green Chem.*, 2013, **15**, 2895–2903.
12. X. Chen, W.-D. Oh, Z.-T. Hu, Y.-M. Sun, R. D. Webster, S.-Z. Li and T.-T. Lim, *Appl. Catal., B*, 2018, **225**, 243–257.
13. J. Li, S. Wang, H.-Y. Liu, H.-J. Zhou and Y. Fu, *ChemistrySelect*, 2017, **2**, 33–41.
14. O. Guellati, F. Antoni, M. Guerioune and D. Bégin, *Catal. Today*, 2018, **301**, 164–171.
15. S. Liu, G. Li, Y. Gao, Z. Xiao, J. Zhang, Q. Wang, X. Zhang and L. Wang, *Catal. Sci. Technol.*, 2017, **7**, 4007.
16. R. Tenne, *Adv. Mater.*, 1995, **7**, 965–995.

17. Y. Marco, L. Roldán, E. Muñoz and E. García-Bordejé, *ChemSusChem*, 2014, 7, 2496–2504.
18. I. Kvande, S. T. Briskeby, M. Tsypkin, M. Rønning, S. Sunde, R. Tunold and D. Chen, *Top. Catal.*, 2007, **45**, 81–85.
19. A. J. Plomp, H. Vuori, A. O. I. Krause, K. P. de Jong and J. H. Bitter, *Appl. Catal.*, 2008, **351**, 9–15.
20. C. Pham-Huu, N. Keller, G. Ehret, L. J. Charbonniere, R. Ziessel and M. J. Ledoux, *J. Mol. Catal. A: Chem.*, 2001, **170**, 55–163.
21. C. Pham-Huu, N. Keller, L. J. Charbonniere, R. Ziesselb and M. J. Ledoux, *Chem. Commun.*, 2000, **19**, 1871–1872.
22. A. Jung, A. Jess, T. Schubert and W. Schutz, *Appl. Catal.*, 2009, **362**, 95–105.
23. M. L. Toebes, Y. Zhang, J. Hájek, T. A. Nijhuis, J. H. Bitter, A. J. van Dillen, D. Y. Murzin, D. C. Koningsberger and K. P. de Jong, *J. Catal.*, 2004, **226**, 215–225.
24. M. L. Toebes, F. F. Prinsloo, J. H. Bitter, A. J. van Dillen and K. P. de Jong, *J. Catal.*, 2003, **214**, 78–87.
25. M. J. Ledoux, R. Viera, C. Phan-Huu and N. Keller, *J. Catal.*, 2003, **216**, 333–342.
26. J. Zhu, M. H. Lu, M. S. Li, J. J. Zhu and Y. H. Shan, *Chin. J. Chem.*, 2011, **29**, 655–660.
27. J. Zhu, M. Lu, M. Li, J. Zhu and Y. Shan, *Mater. Chem. Phys.*, 2012, **132**, 316–323.
28. J. Zhu, M. Li and M. Lu, *Adv. Mater. Res.*, 2012, **535–537**, 178–185.
29. J. Aumo, S. Oksanen, J. P. Mikkola, T. Salmi and D. Y. Murzin, *Catal. Today*, 2005, **102**, 128–132.
30. J. H. Zhou, M. G. Zhang, L. Zhao, P. Li, X. G. Zhou and W. K. Yuan, *Catal. Today*, 2009, **147**, S225–S229.
31. J. Zhou, G. Liu, Z. Sui, X. Zhou and W. Yuan, *Chin. J. Catal.*, 2014, **35**, 692–702.
32. L. Zhao, J. H. Zhou, Z. J. Sui and X. G. Zhou, *Chem. Eng. Sci.*, 2010, **65**, 30–36.
33. Y. Yang, C.-J. Sun, D. E. Brown, L. Zhang, F. Yang, H. Zhao, Y. Wang, X. Ma, X. Zhang and Y. Ren, *Green Chem.*, 2016, **18**, 3558.
34. R. T. K. Baker, K. Laubernds, A. Wootsch and Z. Paal, *J. Catal.*, 2000, **193**, 165–167.
35. V. V. Chesnokov, A. S. Chichkan and Z. R. Ismagilov, *Kinet. Catal.*, 2017, **58**, 649–654.
36. V. V. Chesnokov, O. Y. Podyacheva and R. M. Richards, *Mater. Res. Bull.*, 2017, **88**, 78–84.
37. S. Ordóñez, E. Díaz, R. F. Bueres, E. Asedegbega-Nieto and H. Sastre, *J. Catal.*, 2010, **272**, 158–168.
38. C. Amorim, G. Yuan, P. M. Patterson and M. A. Keane, *J. Catal.*, 2005, **234**, 268–281.
39. J. A. Baeza, L. Calvo, N. Alonso-Morales, F. Heras, S. Eser, J. J. Rodriguez and M. A. Gilarranz, *Chem. Eng. Sci.*, 2018, **176**, 400–408.

40. S. Gryglewicz, A. Śliwak, J. Ćwikła and G. Gryglewicz, *Catal. Lett.*, 2014, **144**, 62.
41. S. Gryglewicz, A. Śliwak, J. Ćwikła and G. Gryglewicz, *Can. J. Chem. Eng.*, 2016, **94**, 1328–1335.
42. F. Salman, C. Park and R. T. K. Baker, *Catal. Today*, 1999, **53**, 385–394.
43. S. Van de Vyver, J. Geboers, M. Dusselier, H. Schepers, T. Vosch, L. Zhang, G. Van Tendeloo, P. A. Jacobs and B. F. Sels, *ChemSusChem*, 2010, **3**, 698–701.
44. S. Van de Vyver, J. Geboers, W. Schutyser, M. Dusselier, P. Eloy, E. Dornez, J. W. Seo, C. M. Courtin, E. M. Gaigneaux, P. A. Jacobs and B. F. Sels, *ChemSusChem*, 2012, **2**, 1549–1558.
45. A. L. Jongerius, R. W. Gosselink, J. Dijkstra, J. H. Bitter, P. C. A. Bruijnincx and B. M. Weckhuysen, *ChemCatChem*, 2013, **5**, 2964–2972.
46. E. Santillan-Jimenez, M. Perdu, R. Pace, T. Morgan and M. Crocker, *Catalysts*, 2015, **5**, 424–441.
47. R. Moreira, E. Ochoa, J. L. Pinilla, A. Portugal and I. Suelves, *Catalysts*, 2018, **8**, 127–141.
48. E. Ochoa, D. Torres, R. Moreira, J. L. Pinilla and I. Suelves, *Appl. Catal., B*, 2018, **239**, 463–474.
49. J. Remón, J. Arauzo, L. García, P. Arcelus-Arrillaga, M. Millan, I. Suelves and J. L. Pinilla, *Fuel Process. Technol.*, 2016, **54**, 178–187.
50. S. A. W. Hollak, R. W. Gosselink, D. S. van Es and J. H. Bitter, *ACS Catal.*, 2013, **3**, 2837–2844.
51. Y. Qin, P. Chen, J. Duan, J. Han, H. Lou, X. Zheng and H. Hong, *RSC Adv.*, 2013, **3**, 17485–17491.
52. R. W. Gosselink, D. R. Stellwagen and J. H. Bitter, *Angew. Chem., Int. Ed.*, 2013, **52**, 5089–5092.
53. D. R. Stellwagen and J. H. Bitter, *Green Chem.*, 2015, **17**, 582–593.
54. B. C. Thompson, E. Murray and G. G. Wallace, *Adv. Mater.*, 2015, **27**, 7563–7582.
55. L. Souza Macedo, R. R. Oliveira Jr., T. Haasterecht, V. Teixeirada Silva and J. H. Bitter, *Appl. Catal., B*, 2018, **241**, 81–88.
56. M. S. Kim, N. M. Rodriguez and R. T. K. J. Baker, *J. Catal.*, 1992, **134**, 253–268.
57. A. Chambers, T. Nemes, N. M. Rodriguez and R. T. K. Baker, *J. Phys. Chem. B*, 1998, **102**, 2251–2258.
58. N. M. Rodriguez, M. S. Kim and R. T. K. Baker, *J. Phys. Chem.*, 1994, **98**, 13108–13111.
59. P. Liu, H. Xie and S. Tan, *React. Kinet. Catal. Lett.*, 2009, **97**, 101.
60. J. L. Pinilla, H. Purón, D. Torres, I. Suelves and M. Millan, *Carbon*, 2015, **81**, 574–586.
61. Z. X. Yu, L. E. Fareid, K. Moljord, E. A. Blekkan, J. C. Walmsley and D. Chen, *Appl. Catal., B*, 2008, **84**, 482–489.
62. Y. N. Prajapati and N. Verma, *Energy Fuels*, 2018, **32**, 2183–2196.

63. J. L. Pinilla, H. Purón, D. Torres, S. de Llobet, R. Moliner, I. Suelves and M. Millan, *Appl. Catal., B*, 2014, **148/149**, 357–365.
64. S. A. AL-Hammadi, A. M. Al-Amer and T. A. Saleh, *Chem. Eng. J.*, 2018, **345**, 242–251.
65. I. T. Ghampson, C. Sepulveda, R. Garcia, J. L. G. Fierro and N. Escalona, *Catal. Sci. Technol.*, 2016, **6**, 4356–4369.
66. M. Breysse, E. Furimsky, S. Kasztelan, M. Lacroix and G. Perot, *Catal. Rev.: Sci. Eng.*, 2002, **44**, 651–735.
67. E. Aryee, A. K. Dalai and J. Adjaye, *Top. Catal.*, 2014, **57**, 796–805.

CHAPTER 9

Catalysts Supported on Graphenes

Structurally, the surfaces of the basal planes of pristine graphene and those of CNTs, CNFs and CNHs are similar. Also, these carbon nanomaterials possess high hydrophobicity and low reactivity. As has been described earlier, these properties can be changed by functionalization and doping. The aspects of graphene and graphene-supported catalysts during HYD and HPR reactions have been discussed in detail elsewhere.[1] Both noble metals and transition metals have been evaluated as active metals. In several studies, graphene-supported catalysts were tested in comparison with catalysts supported on at least one other carbon nanosupport. Therefore, some information on the former catalysts has already been given in the discussions of other carbon nanosupports. Generally, the experimental conditions (*e.g.* reactants, feeds, methodology) used for the evaluation of catalysts supported on carbon nanosupports (*e.g.* CNTs, CNFs and CNHs) are similar to those which have been used for the evaluation of catalysts supported on graphene-derived supports such as graphene oxide (GO) and reduced graphene oxide (rGO).

9.1 Noble Metal-containing Catalysts

The noble metals such as Pt, Pd, Ru and Rh that have been used for the preparation of catalysts supported on CNTs and CNFs have also been used for testing catalysts supported on graphene-derived supports. Similar reactants and feeds to those used for testing of the former catalysts have also been used for the evaluation of graphene-supported catalysts. Most frequently, batch reactor systems have been used.

Catalysis Series No. 37
Carbon Nanomaterials in Hydrogenation Catalysis
By Edward Furimsky

Published by the Royal Society of Chemistry, www.rsc.org

9.1.1 Conversion of α,β-Unsaturated Aldehydes

Few studies have involved a direct comparison of CNT-supported catalysts with graphene supported catalysts. For example, as part of a study of Pt catalysts, an rGO support was compared with CNT and AC supports in the HYD of CALD (313 K, 2 MPa, batch reaction in 2-propanol).[2] Pt(2.9%) catalyst supported on rGO outperformed the other catalysts, as shown in Table 9.1.[2] In spite of excellent textural properties (*e.g.* surface area), the AC support was inferior. This suggests that for highly ordered carbon nanosupports, textural properties play a minor role compared with electronic effects exerted by active metals on carbon supports.[3] A similar absence of these effects may be expected for CB supports.

Pd(5%) catalysts supported on few-layers graphene (FLG) and CNTs were tested for the conversion of CALD in a mixture with dioxane.[4] The experiments were conducted in a batch reactor at 80 °C with bubbling of H_2. The graphene was prepared by the ablation of a graphitic solid followed by sonication and CNTs were prepared by CVD using ethane as the source of carbon over an Fe/Al_2O_3 catalyst. The Pd catalysts were prepared by incipient wetness impregnation using an aqueous solution of $Pd(NO_3)_2{\cdot}6H_2O$. TEM data revealed rather large Pd particles on the CNT supports, ranging from 10 to 50 nm. Two sizes of Pd particles, *i.e.* less than 6 nm and ∼20 nm, were present on the FLG support. The latter particles adopted a flat- instead of a round-shaped morphology. Under the test conditions, the HYD activity of the Pd/FLG catalyst was higher by about 33% than that of the Pd/CNT catalysts, with selectivities with respect to HCALD of 90 and 80%, respectively.

Ru(4%) catalysts supported on either GO or CNTs were compared in the HYD of CALD (batch system, 373 K, 1 MPa H_2) by Wang *et al.*[5] The GO support was prepared by the exfoliation of oxidized graphite and nitric acid-treated commercial CNTs were used for the preparation of the Ru/CNT catalysts. The catalysts were prepared *via* incipient wetness impregnation of the supports using an acetone solution of $RuCl_3{\cdot}3H_2O$.

Based on the results in Table 9.2,[5] the Ru/GO catalysts exhibited better performance than the Ru/CNT catalysts. It was evident that the catalyst performance could be optimized by heat treatment. However, above 700 °C, the heat treatment had an adverse effect on the activity, as indicated by lower TOFs over the Ru/GO-ht and Ru/CNT-ht catalysts compared with the corresponding catalysts that were not heat treated.

Table 9.1 Effect of support type on CALD conversion and selectivity.[2]

Catalyst	Surface area/$m^2\,g^{-1}$	Conversion/%	Selectivity/%				Activity/mol CALD $(mol\,Pt)^{-1}\,h^{-1}$
			CALC	HCALD	HCALC	Others[a]	
Pt/AC	1029	84	30.9	27.0	35.2	6.8	311
Pt/CNT	158	87	48.3	20.8	26.2	4.7	322
Pt/rGO	276	90	69.6	9.2	17.6	3.6	456

[a]Acetals and semiacetals.

Table 9.2 Activity and selectivity in CALD conversion.[5]

Catalyst[a]	TOF[b]/h^{-1}	Selectivity[b]/%			
		CALC	HCALD	HCALC	Others[c]
Ru/GO	119.4	30	55	8	7
Ru/GO-ht	104.6	30	55	7	8
Ru/CNT	52.4	32	52	8	8
Ru/CNT-ht	45.2	39	49	8	4

[a]CO-ht and CNT-ht represent heat-treated catalysts.
[b]TOF and selectivity calculated at 40% conversion.
[c]Acetals and semiacetals.

Table 9.3 Compositions of supports and catalysts and activities of catalysts.[6]

Catalyst	Composition/wt%				Conversion/%	Selectivity/%	
	C	O	N	Pd		HCALD	HCALC
GO	62.5	37.5					
rGO	80.8	19.2					
$NrGO_{150}$	84.4	13.2	2.4				
$NrGO_{300}$	79.9	12.6	7.5				
Pd/rGO	78.1	17.7		4.5	9.0	87.3	12.7
Pd/GO	80.0	15.5		4.6	8.8	86.4	13.6
Pd/$NrGO_{150}$	80.6	12.3	2.5	4.5	51.5	93.4	6.6
Pd/$NrGO_{300}$	75.9	12.2	7.3	4.6	82.5	95.9	4.1

Table 9.3 summarizes the results on the effect of the conditions used in the preparation of rGO from GO by Nie *et al.*[6] The objective of their study was the doping of GO with nitrogen using urea. For this purpose, GO was treated with urea with a ratio of urea to GO of either 300 or 150 to give supports denoted $NrGO_{300}$ and $NrGO_{150}$, respectively. These supports were then used for the preparation of Pd/$NrGO_{300}$ and Pd/$NrGO_{150}$ catalysts. Model reactions such as the HYD of CALD to HCALC (Figure 5.2) as the targeted product and the HYD of phenol to cyclohexanone were used to compare the activities and selectivities of the catalysts. As the results in Table 9.3[6] show, the content of nitrogen in the support had a pronounced effect on catalyst performance, as indicated by both conversion and selectivity. Thus, the study showed that modification of graphene supports with urea nitrogen is an efficient way to improve catalyst performance. The high selectivity of the catalysts for these reactions under mild conditions (*e.g.* 70 °C, 2 MPa of H_2 in a batch reactor) should be noted.

The study by Ji *et al.*[7] is among the few studies in which CB was directly compared with graphene (G) as a support for Pt(3.5%) catalysts. In this case, the G support was prepared using an *in situ* self-generating template method developed by Wang *et al.*[8] It was noted that the oxygen content of this G was much lower than that usually observed for rGO supports. Before catalyst preparation, the G was purified by treatment with HCl to remove Fe. After washing and drying, a slurry of G in deionized water was prepared by sonication. Subsequently, a solution of $H_2PtCl_6{\cdot}6H_2O$ was added to the slurry

Table 9.4 Conversion and product distribution in the HYD of CALD.[7]

		Selectivity/mol%			
Catalyst	Conversion/%	CALC	HCALD	HCALC	Others[a]
Pt/G[b]	73	82	8	6	4
Pt/CB[b]	89	27	10	5	58
Pt/G[c]	92	88	3	7	2
Pt/CB[c]	93	27	9	7	57

[a]Includes 1-(3-propoxyprop-1-enyl)benzene, cinnamyl formate, cinnamic acid, benzyl cinnamate, 4,4-diphenylcyclohexa-1,5-dienyl acetate and others.
[b]Conditions: 50 mg Pt/G, 1.00 g CALD, 30 mL 2-propanol, 1 MPa H_2, 333 K, 2 h.
[c]Conditions: as above except duration 4 h.

and the mixture was sonicated to obtain Pt(3.5%)/G and Pt(3.5%)/CB catalysts. For the latter catalyst, a commercial Vulcan CB was used. The catalysts were applied in the HYD of CALD in a batch reactor (333 K, 1 MPa), the objective being high selectivity with respect to CALC. In this regard, a superior performance of the Pt/G catalyst compared with the Pt/CB catalyst was evident from the results in Table 9.4.[7] It should be noted that the surface area of the Pt/CB catalyst was 198 $m^2 g^{-1}$ compared with 57 $m^2 g^{-1}$ for the Pt/G catalyst. This coincides with the Pt particle size, *i.e.* 3.8 and 4.4 nm, respectively, indicating a higher distribution of Pt on Pt/CB. The significantly greater Raman D/G intensity ratio for the Pt/CB compared with the Pt/G catalyst, *i.e.* 1.2 and 0.3, respectively, provides clear confirmation of the amorphous structure of the CB. Therefore, electronic effects of the support played a much more important role than textural properties in the catalyst selectivity.

9.1.2 Conversion of Lignocellulosic Reactants

The products and intermediates observed in the depolymerization of cellulose and hemicellulose (Figure 5.1) and also those of lignin (Figure 5.5) have been investigated as reactants for the evaluation of the activity and selectivity of catalysts supported on graphene supports. For example, the gradual conversion of cellulose *via* cellobiose (dimer of glucose) to glucose and sorbitol was studied by Wang *et al.*[9] over Pt(5%)/rGO catalysts. The experiments were conducted in an autoclave at 420–480 K with 5 MPa of H_2. For comparison, supports such as graphite, ACs, GO, CNTs and SiO_2 were also used. For the preparation of the Pt/rGO catalyst, the support and H_2PtCl_6 were dispersed ultrasonically in ethylene glycol, then the mixture was microwaved at 393–453 K. The solid obtained was filtered of and washed with water and ethanol. Pt catalysts supported on the other supports were prepared in a similar manner and Pt nanoparticles were prepared using the same method in the absence of the support. The objective of the study was to maximize the yield of sorbitol. During the conversion of cellulose, the following order of selectivity with respect to sorbitol was established: Pt/rGO > Pt/CNT > Pt nanoparticles > Pt/AC > Pt/SiO_2 > Pt/graphite. The effect

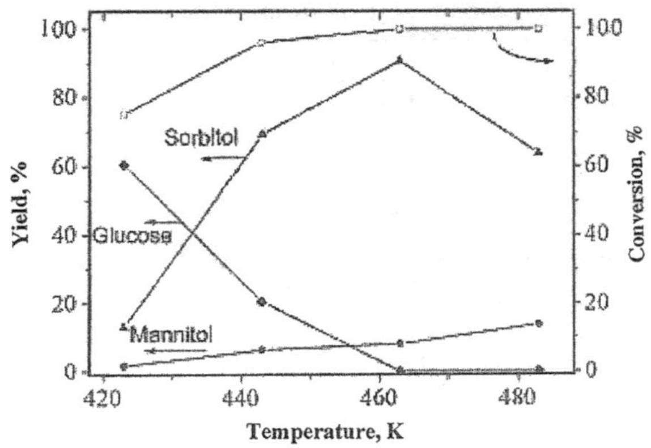

Figure 9.1 Effect of temperature on the product distribution in the conversion of cellobiose over Pt/rGO.
Reproduced from ref. 9 with permission from John Wiley and Sons, © 2014 WILEY-VCH Verlag GmbH & Co. KGaA, Weinheim.

of temperature on the product distribution in the conversion of cellobiose over the most selective Pt/rGO catalyst is shown in Figure 9.1.[9] It is evident that temperature is an important parameter for optimizing the production of sorbitol. In the same study, the metals Ni, Cu, Ru, Rh, Pd and Pt were compared using rGO as the best support. In this case, the catalysts contained 5 wt% of metals. In this series of catalysts, the best performance was exhibited by the Pt/rGO catalyst. On the basis of these results, in the conversion of cellobiose at 463 K with 5 MPa of H_2 the following order of the selectivity with respect to sorbitol could be established: Pt/rGO > Ru/rGO > Rh/rGO > Ni/rGO > Pd/rGO > Cu/rGO. For the most selective Pt/rGO catalyst, the yield of sorbitol exceeded 90%. When cellulose was used as the feed, the yield of sorbitol exceeded 60%. Figure 9.2[9] shows that the performance of the catalyst can be further improved during catalyst preparation by optimizing the size of the Pt particles.

Ramirez-Barria *et al.*[10] prepared GO and rGO by the exfoliation of graphite oxide under either N_2 or NH_3 to obtain GO and NrGO, respectively. These solids were used as supports for the preparation of Ru(4%) catalysts *via* wetness impregnation using $Ru_3(CO)_{12}$ dissolved in acetone to obtain the catalysts denoted with the suffix CO. Another two series of catalysts were prepared using a water–ethanol solution of either $RuCl_3$ or $Ru(NO)(NO_3)_3$ and were denoted with the suffixes Cl and NN, respectively. The catalysts were activated in a flow of H_2 at either 300 or at 350 °C with the aim of decomposing the Ru precursor. For comparison, Ru(4%) catalysts supported on AC, SiO_2, TiO_2, Al_2O_3 and high surface area graphite (G) were prepared also using $Ru(NO)(NO_3)_3$ precursor. The catalysts were used for the conversion of furfural (FF) to furfuryl alcohol (FFA) in a batch reactor at 20 °C and 1 MPa of H_2. The results of these evaluations, including conversions, selectivities with respect to FFA and THFA and TOFs, are presented in

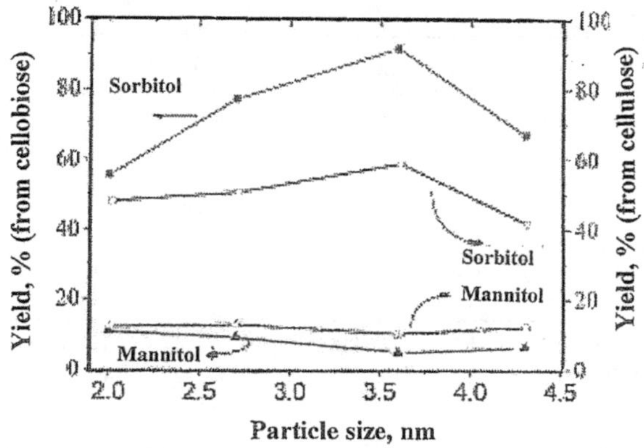

Figure 9.2 Effect of the size of Pt nanoparticles on the product distribution in the conversion of cellobiose over Pt/rGO.
Reproduced from ref. 9 with permission from John Wiley and Sons, © 2014 WILEY-VCH Verlag GmbH & Co. KGaA, Weinheim.

Table 9.5 Conversion of furfural (FF) to furfuryl alcohol (FFA) and tetrahydroFFA (THFFA) over Ru catalysts prereduced at 350 °C.[10]

Catalyst[a]	Conversion/%	Selectivity/%		TOF/h^{-1}
		FFA	THFFA	
rGO	3.0	100	0	—
Ru-Cl/rGO	49	98	2	5.0
Ru-NN/rGO	73	98	2	7.9
Ru-NN/rGO[a]	93	97	3	9.0
Ru-CO/rGO	83	98	2	8.6
Ru-CO/rGO[b]	93	98	2	9.6
Ru-CO/G	46	99	1	7.8
Ru-CO/G[a]	50	98	2	6.6
Ru-CO/AC	51	99	1	6.0
Ru-CO/Al_2O_3	28	100	0	2.0
Ru-CO/SiO_2	30	100	0	5.2
Ru-CO/TiO_2	25	100	0	2.6

[a]Conditions: 20 °C, 1 MPa H_2, 5 h, solvent water.
[b]Catalysts prereduced at 300 °C.

Table 9.5.[10] First, in terms of activity, the catalysts supported on rGO outperformed the other catalysts. Among the rGO-supported catalysts, those prepared using $Ru_3(CO)_{12}$ and $Ru(NO)(NO_3)_3$ precursors exhibited better performance than that prepared using $RuCl_3$ precursor. For the catalyst prepared with $Ru(NO)(NO_3)_3$ precursor, a further improvement in activity was achieved by prereduction of the catalyst at 300 °C compared with that at 350 °C. Carbon supports such as AC and G were much less suitable than the rGO support. Even more inferior were oxidic supports; because of the aqueous phase employed, the poor performance of the catalysts supported on oxidic supports is not surprising.

Table 9.6 Activities and selectivities of catalysts for the HYD of 5-HMF.[11]

Catalyst	Pt content/wt%	Pt particle size/nm	Conversion[a]/% DMF	MFF	BHMF	MFM	Yield/% of carbon
rGO	—	0	10.5	0	2.9	1.4	1.0
PtNPs[a]	—	4.9	35.6	19.9	1.0	0.9	1.1
Pt/rGO	4.9	2.8	100	73.2	1.6	0.8	0.6
Pt/AC	4.5	3.3	67.4	32.6	2.5	2.8	0
Pt/SiO_2	3.9	4.6	52.1	18.2	0	0	2.2
Pt/Al_2O_3	4.1	5.7	45.2	16.4	0.9	0	1.0

[a]DMF, 2,5-dimethylfuran; MFF, 5-methylfurfural; BHMF, 2,5-bishydroxymethylfuran; MFM, 5-methyl-2-furanmethanol.

A study by Shi *et al.*[11] provides evidence for the superior performance of Pt catalysts supported on rGO in comparison with those supported on SiO_2 and Al_2O_3. In this case, the HYD of 5-hydroxymethylfurfural (5-HMF) to 2,5-dimethylfuran (DMF) was used as a model reaction. Interest in this product as a suitable blending component of gasoline is due to its high octane number. The results of the comparison of these catalysts, summarized in Table 9.6,[11] indicate that the Pt/rGO catalyst showed the best performance, particularly in terms of the yield of desirable products such as DMF. Spectroscopic data on the dispersion of Pt metal on the supports identified rGO as the most suitable support. The mild conditions used for the evaluation of the catalyst performance (*e.g.* 120 °C and 3 MPa of H_2 in a batch reactor) should be noted.

In order to develop Brønsted acidity, GO was sulfonated before being used as a support for the preparation of catalysts consisting of a multifunctional Pd/Zr-based metal–organic framework.[12] This catalyst was used in the first step of the overall conversion of fructose to 2,5-DMF. In this case, fructose was first converted to 5-HMF over the catalyst, followed by a second step involving HYD and hydrogenolysis of 5-HMF to 2,5-DMF over a Pd catalyst. The overall conversion to 2,5-DMF exceeded 70% under fairly mild conditions (*e.g.* 160 °C, 1 MPa of H_2, batch system). The catalyst exhibited high activity and selectivity in several repeated cycles.

Nanocomposites consisting of Ce-based metal–organic framework crystals and partially reduced GO nanosheets were used as the support for Pd nanoparticles by Ibrahim *et al.*[13] The resulting catalyst was applied in the HDO of vanillin in an aqueous solution. The catalyst exhibited a high HDO activity in spite of the mild conditions employed (*e.g.* 100 °C and 1 MPa of H_2 in a batch reactor). The spent catalyst was separated from the reaction mixture, washed with DMF and ethanol, then dried at 100 °C. After several subsequent cycles, the activity of the spent-reactivated catalyst was still similar to that of the corresponding fresh catalyst.

Ru/GO and Ru/AC catalysts were compared in the conversion of levulinic acid (LA) to γ-valerolactone (GVL) by Upare *et al.*[14] The catalysts were prepared by soaking either GO or AC in an aqueous solution of $RuCl_3$ and subsequent sonication for 1 h followed by aging for 12 h at room temperature. The powder obtained after evaporation of water was dried and reduced in a flow of H_2–N_2 at 450 °C. The experiments were conducted in a continuous fixed-bed system (265 °C, 1–2.5 MPa of H_2) using a mixture of GVL (10%) in 1,4-dioxane. Over Ru/GO, the yield of GVL was 41% and cyclic ethers such as 2-methyltetrahydrofuran and THF combined accounted for about 54% of the converted LA. Only GVL was formed over the Ru/AC catalyst at a much lower overall conversion of LA than that over the Ru/GO catalyst.

The effect of sulfonation on the properties of rGO as the support of Ru catalysts was investigated by Wang *et al.*[15] The objective of sulfonation was to develop Brønsted acidity *via* the introduction of a sulfonic group such as $-SO_3H$. The Ru catalysts supported either on sulfonated rGO (rGO-s) or on unsulfonated rGO were compared in the conversion of LA in an aqueous solution (50 °C, 2 MPa of H_2, batch system). The objective was a high yield of GVL. According to Figure 9.3,[1,15] the selectivity with respect to GVL over the Ru/rGO-s catalyst was significantly greater than that over the Ru/rGO catalyst. Apparently, the Brønsted acidity supplied by sulfonic groups on the rGO-s support was responsible for the high selectivity. In this case, the dehydration of the 4-hydroxyvaleric acid intermediate to GVL was aided by protons provided by the sulfonic groups. The intermediate was formed in the first step of the overall conversion of LA. This study[15] confirmed that

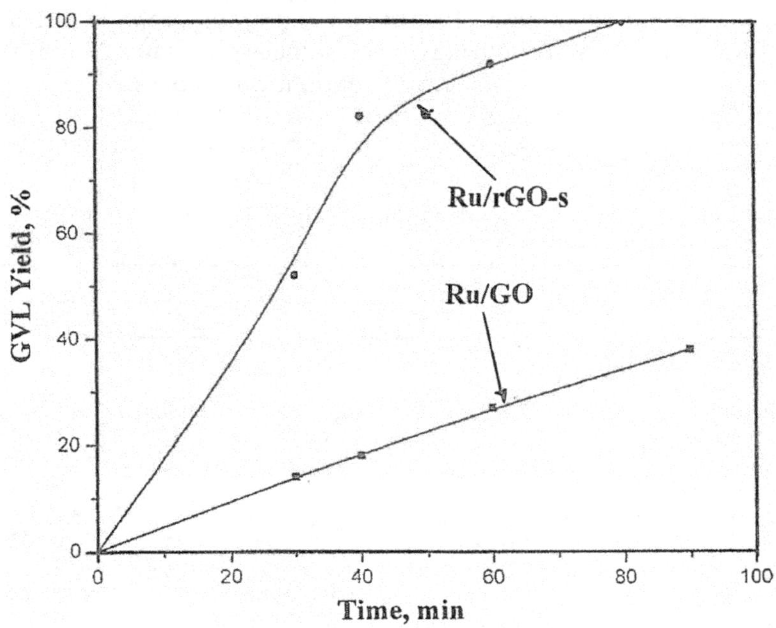

Figure 9.3 Effect of sulfonation of the rGO support on the activity of an Ru/rGO catalyst.
Reproduced from ref. 15 with permission from American Chemical Society, Copyright 2017.

sulfonation appears to be a suitable method for developing a bifunctional nature of catalysts supported on carbon nanosupports.

Ru catalysts supported on few-layers graphene (FLG), ACs and CBs were compared with a commercial Ru/AC catalyst in the conversion of LA to GVL in a batch reactor at room temperature and 2 MPa of H_2.[16] The FLG was obtained by the thermal exfoliation of graphite oxide at 700 °C in a flow of H_2–Ar mixture. The Ru nanoparticles were prepared by the reduction of $RuCl_3$ in ethylene glycol in the presence of sodium hydroxide. After separation, Ru nanoparticles were dispersed in ethanol and mixed with FLG. On the basis of conversion and selectivity, the Ru/FLG catalyst outperformed Ru/AC, Ru/CB and Ru/SiO_2 catalysts and also the commercial catalyst. The optimum performance of the Ru/FLG catalyst was maintained during recycling of the spent Ru/FLG catalysts; the conversion and selectivity after several cycles were 99 and 98%, respectively.

9.1.3 Conversion of Aromatics

Noble metal catalysts supported on carbon nanosupports have been evaluated for potential applications in the upgrading of fuels to meet regulatory limits on aromatics. For example, Yao *et al.*[17] used benzene and toluene as model compounds to determine the activity of Ru/rGO catalysts in comparison with Ru/MCF-C and Ru/CAC catalysts, where the MCF-C and CAC supports are a siliceous mesocellular foam and an amorphous carbon, respectively. Testing was conducted in an autoclave at 130 °C and an H_2 pressure of 4 MPa using a mixture of reactant and *n*-decane. The results (Table 9.7[17]) confirmed the superiority of the Ru/rGO catalyst. For example, at the end of the test (after 60 min), the HYD conversion of toluene over

Table 9.7 Effect of carbon support on the activity of Ru catalysts for the HYD of toluene and benzene.[17]

	Time/min			
Catalyst[a]	15	30	45	60
Toluene/Ru $= 8.8\times10^3$				
Ru/rGO				
Fresh	39.1	88.8	99.8	100
After 6 cycles	38.5	86.5	98.2	100
Ru/MCF-C	33.6	41.8	44.7	45.1
Toluene/Ru $= 5.3\times10^4$				
Ru/rGO	11.6	18.9	25.5	34.3
Ru/CAC	5.8	7.7	9.0	12.6
Ru/MCF-C	4.6	9.6	15.0	21.0
Benzene/Ru $= 5.3\times10^4$				
Ru/rGO	14.5	30.2	44.7	61.3
Ru/MCF-C	6.1	10.0	14.2	19.8

[a]Conditions: autoclave, 130 °C, 4 MPa H_2.

Ru/rGO and Ru/MCF-C catalysts was 34.3 and 21%, respectively, and in the HYD of benzene 61.3 and 19.8%, respectively. It was evident that the surfaces of these catalysts were different. This was confirmed by the results in Figure 9.4,[17] showing the effect of temperature on the desorption of toluene from the catalyst surface. For the Ru/rGO catalyst, the temperature for maximum toluene desorption was about 80 °C higher than that for the Ru/MCF-C catalyst. This can be interpreted in terms of a more efficient interaction of toluene with active sites on the former catalyst. Among the three catalysts tested, the Ru/CAC catalyst was the least active. The high recyclability of the Ru/rGO catalyst should be noted. Thus, after six cycles, complete recovery of activity of spent catalyst was achieved, as shown in Table 9.7.[17] In this case, the catalyst was separated from the reaction mixture by centrifugation, washed with acetone and toluene and dried before being used in the next reaction run conducted under the same conditions. Over Ru/rGO, negligible loss of activity was also observed after several repeated cycles of the HYD of benzene.[18]

Carbon nanomaterials such as functionalized GO and CNTs were compared as supports for Ru catalysts by Guo *et al.*[19] The preparation of the catalysts involved incipient wetness impregnation of the supports. Catalysts heat treated at 700 °C, *i.e.* Ru/GO-ht and Ru/CNT-ht, were also tested. In these catalysts, the size of the Ru nanoparticles was 2.3 nm. The activity of the catalysts was evaluated in the HYD of benzene and *p*-chloronitrobenzene (*p*-CNB) as model compounds. The experiments were carried out in an autoclave at 70 °C and 1 MPa of H_2. The results of these evaluations are given

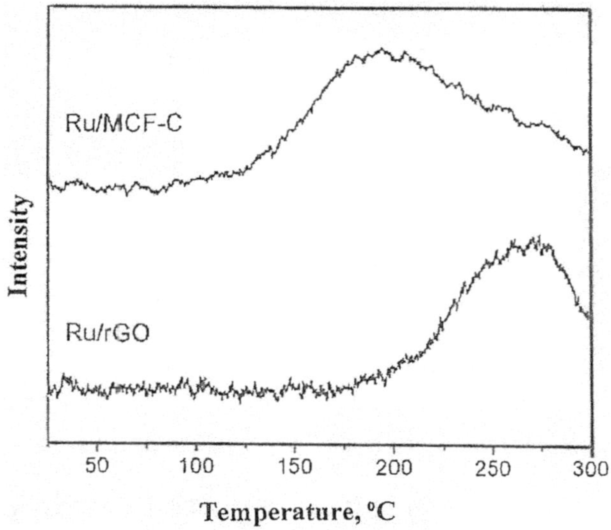

Figure 9.4 Effect of catalyst type on adsorption of toluene.
Reproduced from ref. 17 with permission from John Wiley and Sons, Copyright © 2012 WILEY-VCH Verlag GmbH & Co. KGaA, Weinheim.

Table 9.8 Catalyst performance in the conversion of benzene and *p*-CNB.[19]

Reactant	Catalyst	Conversion[a]/%	TOF[b]/h^{-1}
Benzene	Ru/GO	74.5	1302
	Ru/GO-ht	23.5	411
	Ru/CNT	51.2	622
	Ru/CNT-ht	24.7	375
p-CNB	Ru/GO	5.3	93
	Ru/GO-ht	9.8	172
	Ru/CNT	14.3	174
	Ru/CNT-ht	32.5	494

[a]Conversion was recorded at 0.5 h for all data.
[b]TOF: mol reactant (mol Ru)$^{-1}$ h^{-1}.

in Table 9.8.[19] The higher activity of the Ru/GO catalyst for the HYD of benzene was in contrast with higher activity of the Ru/CNT catalyst for the HYD of *p*-CNB. This was attributed to the more efficient electron-donating ability of benzene caused by favorable adsorption on the GO support compared with that of *p*-CNB.

Karakulina *et al.*[20] prepared an Rh/rGO catalyst by the deposition of Rh on rGO supports using $Rh_6(CO)_{16}$ precursor. The catalyst was applied in the partial HYD of quinoline to tetrahydroquinoline in a batch system at 80 °C with H_2 pressure varying from 1 to 3 MPa. For the tests, powdered Rh/rGO and reactant were slurried with an ionic liquid such as [bmim][BF_4] (bmim = 1-butyl-3-methylimidazolium cation). Under these conditions, no other products were formed as the selectivity with respect to tetrahydroquinoline was 99%.

9.1.4 Conversion of Chlorinated Reactants

The Pd/GO catalyst exhibited high activity in the hydrodechlorination (HDCl) of 4-chlorophenol and 2,3,5-trichlorophenol.[21] Thus, in a batch reactor, complete conversion of these reactants was achieved over a Pd(3%)/GO catalyst in the presence of K_2CO_3 or Na_2CO_3 at room temperature and atmospheric pressure H_2 within 2 h. However, over Pd(1%)/GO, the decomposition took a longer time. After six recycles, the Pd/GO catalyst retained almost all of its original activity. An Ni–Pd catalyst supported on nitrogen-doped graphene exhibited high activity in the HDCl of dichloroaromatics to aromatics.[19] The experiments were conducted in aqueous 2-propanol at 50 °C in a batch reactor using NH_3BH_3 instead of H_2 as reducing agent.

9.2 Conventional Metals Supported on Graphene

Transition metals such as Co, Ni, Mo, W, Cu and Fe have been essential active metals in conventional HPR catalysts.[22–24] Similarly to other carbon nanosupports, Mo_2C and W_2C have been tested as active phases.

9.2.1 α,β-Unsaturated Aldehydes

Mohire and Yadev[25] prepared monometallic Ni and Cu catalysts and bimetallic Ni–Cu catalysts supported on rGO with various Ni : Cu ratios (Table 9.9) and applied them in the conversion of CALD in a batch reactor at 150 °C and 2 MPa using methanol as solvent. The objective was to maximize the production of HCALD. Table 9.9[25] shows that the bimetallic catalysts exhibited higher activity and selectivity. The catalyst performance could be further optimized by varying the amount of active metals. To determine the stability, spent Ni(5%)–Cu(1 : 1)/rGO catalyst was recovered from the reaction mixture by filtration, washed with water and ethanol, vacuum dried at 60 °C and then recycled in another run. Little activity loss was observed after five repeated runs. The high catalyst stability was attributed to a negligible loss of active metals due to leaching.

9.2.2 Lignocellulosic Reactants

A novel approach to catalyst preparation was presented by Chen *et al.*[26] Monometallic Co and Ni and bimetallic Co–Ni catalysts were prepared by coating the active metals with several layers of graphene. The modified Pechini sol–gel method[27] was used for catalyst preparation. The catalysts were applied in the conversion of lignin model reactants such as 2-phenoxy-1-phenylethanol. This reactant has been used to study lignin depolymerization *via* hydrogenolysis of the β-O-4 linkages, which account for about 50% of all linkages in lignin.[28] For testing, the reactant was mixed with ethanol–water solvent. The tests were performed in a batch reactor at 170 °C and 2 MPa of H_2 for 6 h. For the Co catalyst, the conversion involved the HDO of hydroxyl groups with little hydrogenolysis of etheric bridges. On the other hand, over the Ni catalyst, hydrogenolysis of the etheric bond was quite evident. The best performance in terms of hydrogenolysis selectivity was observed over the bimetallic Co–Ni catalyst, *e.g.* for the catalyst with an Ni : Co ratio of 1 : 3, more than 80% selectivity at complete conversion was achieved. High stability of the catalyst was demonstrated by little change in activity and selectivity after recycling the catalyst eight times. A decrease in activity caused by catalyst deactivation was observed in the ninth run.

Table 9.9 Conversion of CALD to HCALD over Ni–Cu catalysts.[25]

Catalyst	Conversion/%	Selectivity/%
Cu(5%)/rGO	66	14
Ni(5%)/rGO	52	28
Ni(2%)–Cu(1 : 1)/rGO	70	100
Ni(5%)–Cu(1 : 1)/rGO	82	100
Ni(10%)–Cu(1 : 1)/rGO	89	100
Ni(5%)–Cu(2 : 1)/rGO	56	57
Ni(5%)–Cu(3 : 1)/rGO	38	48

Table 9.10 Product distribution in the HDO of 4-methylphenol in the absence and presence of water.[29]

Catalyst[a]	Product[b]/%					DEOX/%
	4-MCHol	4-MCHon	3-MCHen	MCHan	TOL	
NiBPO	40.7	5.7	0.4	52.6	0.7	55.5
NiBPO[a]	64.5	9.3	0.2	24.9	1.1	29.7
NiBPO/rGO	0.3	0.4	0.0	99.9	0.1	99.9
NiBPO/rGO[c]	0	0	0.9	93.2	5.9	99.2
CoBPO	30.4	4.1	0.2	60.3	5.8	66.4
CoBPO[a]	40.4	4.0	0.1	53.3	2.2	56.3
CoBPO/rGO	0.2	0.1	0.3	96.1	3.3	99.2
CoBPO/rGO[a]	0	0	1.3	91.8	6.9	98.9

[a]Conditions: 0.1 g catalyst, 4.8 g phenol in 28.5 g of dodecane, 225 °C, 4 MPa H_2, 1 h.
[b]4-MCHol, 4-methylcyclohexanol; 4-MCHon, 4-methylcyclohexanone; 3-MCHen, 3-methylcyclohexene; MCHan, methylcyclohexane; TOL, toluene; DEOX, degree of deoxygenation.
[c]Water : phenol = 3 : 8.

The generally observed high stability of catalysts supported on carbon supports under aqueous conditions has been attributed to the hydrophobic surface of the carbon. To confirm these effects, the performance of catalysts in the absence and presence of water was investigated by Wang *et al.*[29] The catalysts they prepared consisted of the oxygenated form of Ni and Co boride/phosphide phases (NiBPO and CoBPO). In addition, catalysts comprising these phases but supported on rGO were used for comparison. The performance of these catalysts was evaluated in the HDO of 4-methylphenol in a batch reactor. According to Table 9.10,[29] the activity of the catalysts supported on rGO was much higher than that of the corresponding unsupported catalysts. For the former catalysts, almost complete conversion of 4-methylphenol to methylcyclohexane was achieved over both types of catalyst. The NiBPO/rGO catalyst was more active than the CoBPO/rGO catalyst. This may result from a higher HYD activity of Ni than that of Co, as generally reported. For both catalysts, water had a minor effect on the overall HDO but had an adverse effect on the HYD pathway of the overall 4-methylphenol conversion.

9.2.3 Vegetable Oils and Fatty Acids

Kim *et al.*[30a] used Mo_2C/rGO catalysts for the HDO of oleic acid and soybean oil in a continuous system at 350 °C and 5 MPa of H_2 with the LHSV of the reactant varying from 2.0 to 8.0 h^{-1}. Other carbon supports, *i.e.* AC, a spherical carbon (SC) and mesoporous carbon (MC), were used for comparison together with a sulfided Co–Mo/Al_2O_3 catalyst. Prior to MoO_2 deposition, the carbon supports were treated with nitric acid. MoO_2 was converted into Mo_2C in the carburization step in H_2. To identify the optimal composition, catalysts with Mo loadings varying from 8 to 38 wt% were evaluated. In terms of activity and selectivity with respect to hydrocarbons, the best performance was observed for the catalysts with a 17 wt% Mo_2C

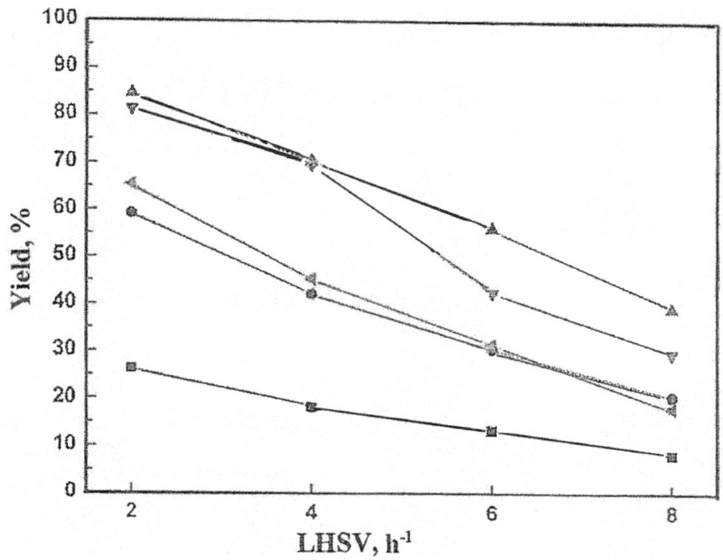

Figure 9.5 Effect of Mo_2C loading on rGO on the yield of hydrocarbons in the conversion of oleic acid: ■, 8%; ●, 12%; ▲, 17%; ▼, 20%; ◀, 38%. Reproduced from ref. 30*a* with permission from American Chemical Society, Copyright 2015.

loading (Figure 9.5).[30a] At this Mo_2C loading, the rGO support was compared with AC, MC and SC supports. Figure 9.6[30a] confirms the evident advantage of rGO over the other supports. This catalyst was compared with the conventional sulfided Co–Mo/Al_2O_3 catalyst in the HDO of soybean oil. As Figure 9.7[30a] shows, the Mo_2C(17%)/rGO catalyst exhibited higher stability than the Co–Mo/Al_2O_3 catalyst with increasing time on-stream. The stability comparison revealed that after 6 h on-stream, the Mo_2C/rGO catalyst had lost ~13% of its activity compared with 43% for the CoMo/Al_2O_3 catalyst. Spectroscopic evaluations revealed that in the Mo_2C/rGO catalysts, Mo_2C was the active phase rather than MoO_2.

9.2.4 Petroleum Reactants and Feeds

Xu *et al.*[30b] investigated the effect of the hexamethonium bromide : Mo ratio in the preparation of CoMoS/GO catalysts. The results were compared with those for the corresponding catalysts prepared by the conventional method. The HDS of thiophene was used as a model reaction. The testing was conducted in a continuous fixed-bed reactor at 280 °C and 1.5 MPa of H_2. The catalyst prepared by the novel method exhibited higher activity and HDS selectivity This was attributed to the finely tuned morphology of the CoMoS active phase, possessing an optimal ratio of the edge to corner Co atoms.

Studies in which the performances of conventional catalysts were evaluated together with novel catalysts supported on carbon nanosupports are of

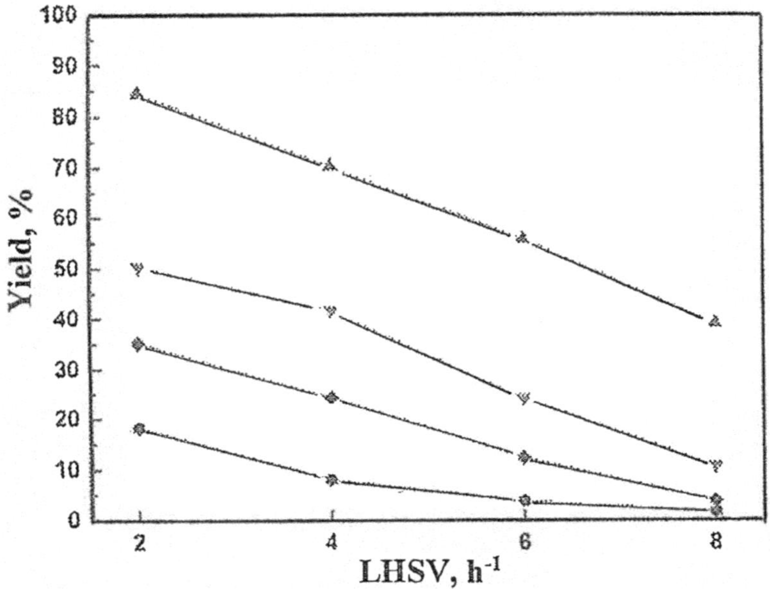

Figure 9.6 Effect of support type and LHSV on the yield of hydrocarbons in the conversion of oleic acid over Mo_2C(17%) catalysts: ▲, rGO; ▼, AC; ♦, MC; ●, SC.
Reproduced from ref. 30*a* with permission from American Chemical Society, Copyright 2015.

particular interest to confirm the superiority of the latter catalysts. In such a study by Wang *et al.*,[31] a composite consisting of graphene-covered mesoporous TiO_2 was compared with TiO_2 alone as supports for Ni–Mo catalysts. The preparation of the composite involved sonication of GO suspended in distilled water–ethanol medium. Subsequently, TiO_2 was added to the suspension and the mixture was transferred to a Teflon-sealed autoclave and maintained at 120 °C. After washing and drying, the composite obtained ($TiGC_x$) was used for catalyst preparation using incipient wetness impregnation with solutions of nickel nitrate hexahydrate and ammonium heptamolybdate. The solid obtained was calcined at 500 °C and sulfided in a CS_2–H_2 mixture to obtain Ni–Mo/$TiGC_x$ catalysts. The activities of the catalysts were determined in the HDS of DBT in a continuous fixed-bed reactor. In terms of the overall HDS conversion, the Ni–Mo/$TiGC_x$ catalysts were more active than the unmodified Ni–Mo/TiO_2 catalyst (Figure 9.8[31]).

During catalyst preparation, attempts are always made to achieve efficient dispersion of the active metal on the surface of the support. This prevents the agglomeration of metal particles and leaching of metals during the catalyst evaluations. It was reported by Wei *et al.*[32] that this can be achieved by encapsulation of active metals between graphene layers. In this case, nitrogen-doped graphene was used as the support and Co nanoparticles as the active metal. The catalyst was applied in the HYD of quinoline to

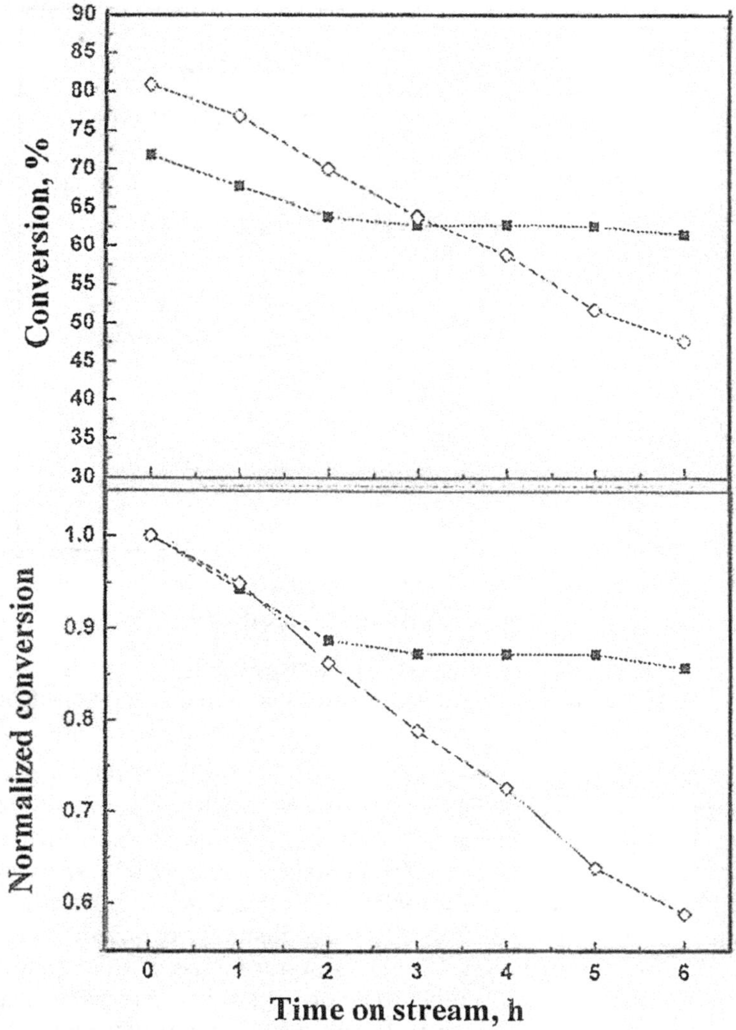

Figure 9.7 Conversion of soybean oil over (■) Mo_2C(17%)/rGO and (◊) CoMoS/Al_2O_3 catalysts.
Reproduced from ref. 30*a* with permission from American Chemical Society, Copyright 2015.

tetrahydroquinoline in methanol. After separation from the reaction mixture, the spent catalyst was washed and dried before being used in further experiments. The method of preparation ensured the stability of the catalyst as confirmed by the almost complete recovery of catalyst activity.

γ-Alumina, as the most frequently used support for the preparation of industrial HPR catalysts, was compared with functionalized graphene as part of Co–Mo catalysts in a study by Hajjar *et al.*[33] The graphene supports, prepared by CVD using either camphor or methane as the source of carbon,

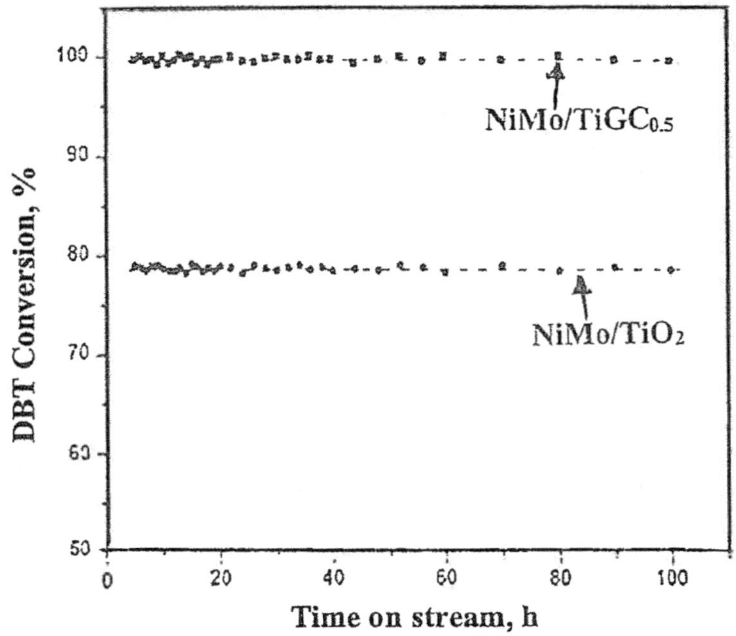

Figure 9.8 Effect of catalyst type on the HDS of DBT.
Reproduced from ref. 31 with permission from Elsevier, Copyright 2015.

were functionalized using a mixture of concentrated HNO_3 and H_2SO_4. These supports were co-currently impregnated with aqueous solutions of ammonium heptamolybdate and Co nitrate. After separation, drying and calcining, the solid obtained was sulfided using a mixture of 1 wt% dimethyl disulfide (DMDS) dissolved in a hydrocarbon fraction. In parallel, γ-Al_2O_3 was impregnated and treated in a similar fashion to obtain the corresponding Co–Mo/Al_2O_3 catalyst. The HDS performance of the catalysts was evaluated in a continuous fixed-bed reactor system using real feeds such as naphtha and diesel fractions derived from conventional petroleum. The results of these tests recorded after 120 h on-stream are summarized in Table 9.11.[33] They indicate superiority of the Co–Mo catalyst supported on camphoric graphene. The same is also evident from the results in Figure 9.9.[33]

In another study by Hajjar *et al.*,[34] the conventional Co–Mo/Al_2O_3 catalyst was compared with Co–Mo catalysts supported on GO. For the latter catalysts, the preparation involved simultaneous exfoliation of graphite oxide and MoS_2 powder. The catalysts were compared in the HDS of naphtha feed containing 1350 ppm of sulfur. The superior performance of the GO-supported catalysts was confirmed by the complete removal of sulfur from the feed. Under the same conditions over the conventional Co–Mo/Al_2O_3 catalyst, about 100 ppm of sulfur was still present in the product obtained. In addition, Hajjar and co-workers[35,36] prepared Co–Mo/GO catalysts *via*

Table 9.11 Sulfur content in products from the HDS of naphtha (1350 ppm sulfur) and diesel (13 000 ppm sulfur).[33]

	Naphtha		Diesel	
Catalyst[a]	S in product/ppm	S removed/%	S in product/ppm	S removed/%
Co–Mo/G-camphoric	0	100	15	99.9
Co–Mo/G-methanic	20	99	21	99.8
Co–Mo/Al_2O_3	100	93	1937	85.1

[a]Conditions: 300 °C, 1.5 MPa of H_2, continuous system.

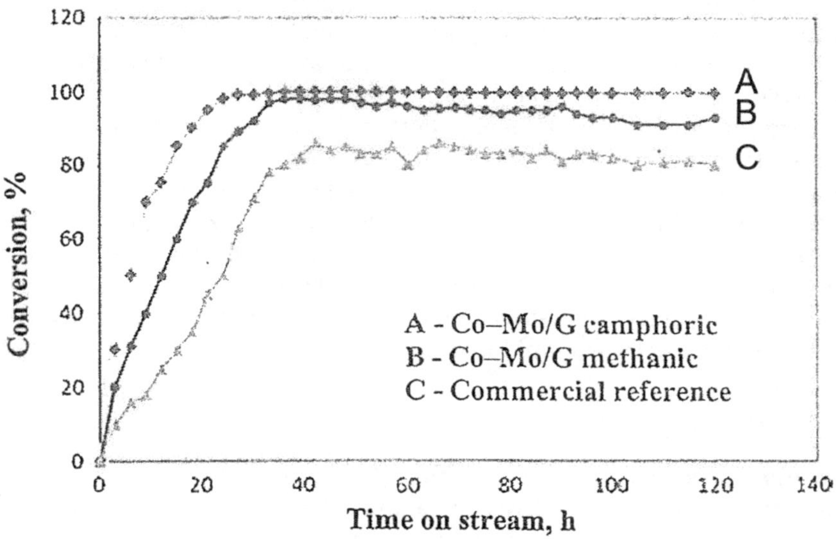

Figure 9.9 Effect of graphene support type on the HDS activity of Co–Mo catalysts. Reproduced from ref. 33 with permission from Elsevier, Copyright 2014.

three different methods, *i.e.* hydrothermal, incipient wetness impregnation and spray pyrolysis. The catalysts were applied in the HDS of naphtha containing 2800 ppm of sulfur conducted under the same conditions. Prior to testing, the catalysts were presulfided at 310 °C and 3 MPa of H_2 for 12 h using a mixture of DMDS (1%) in a hydrocarbon liquid. Commercial Co–Mo/Al_2O_3 catalyst, presulfided and tested under the same conditions, was used for comparison. Almost complete sulfur removal was observed over the GO-supported catalysts compared with about 90% removal over the commercial catalyst.

Ni–Mo catalysts supported on GO, prepared by functionalizing graphene using an H_2SO_4–HNO_3 mixture, were evaluated in the HDS and HDN of a heavy naphtha by Behnejad *et al.*[37] After the incipient wetness impregnation of GO with Mo precursor first, followed by heating at 120 °C and calcination at 350 °C in Ar, Ni was introduced using a similar procedure. The testing was conducted in a continuous fixed-bed reactor at 290 °C, 3 MPa and an LHSV of 3.3 h^{-1}. Under the same conditions, a conventional Ni–Mo/Al_2O_3 catalyst was evaluated for comparison. Both catalysts were used in a sulfided form. For both sulfur and nitrogen removal, the Ni–Mo/GO catalyst outperformed the Ni–Mo/Al_2O_3 catalyst.

In an unusual study by Yuan *et al.*,[38] an extra heavy feed was treated with an Ni–graphene composite with the aim of achieving a decrease in viscosity to simplify downstream handling of the feed. The composite was prepared by magnetic stirring of a mixture comprising 2.00 g of $NiCl_2{\cdot}6H_2O$ and 0.24 g of $NaBH_4$ with 50 mL of ethylene glycol, then sonication of the mixture with rGO at 120–180 °C. The experiments were conducted in a batch reactor at 280 °C. In this case, 50 g of heavy feed mixed with 1 wt% of the nanocomposite and 3 wt% of tetralin were added and mixed in the reactor. Under these conditions, a decrease in viscosity of more than 80% was achieved. It was evident that the Ni–rGO composite was involved in an efficient transfer of hydrogen from tetralin to asphaltenic molecules of the heavy feed.

In another study on viscosity reduction by Xing *et al.*,[39] the focus was on a Chinese heavy feed. The Fe–graphene composite used as the catalyst was prepared according to the methodology developed by Zhang *et al.*[40] Under the same conditions, the decrease in viscosity in the presence and absence of the catalyst was 64 and 38%, respectively. The mild conditions needed to achieve such a decrease in viscosity, *i.e.* 100 g of feed with 1 g of catalyst at 200 °C in an autoclave, should be noted.

9.2.5 Other Reactants

The unusual reactant carbonyl sulfide (COS) has attracted attention as a model feed to study the activity of catalysts for HDS reactions. Liu *et al.*[41] used a gaseous feed that consisted of 500 ppm of COS in a mixture of H_2–N_2 (10 : 90) to study the effect of the conditions employed during preparation on catalyst activity. For this purpose, MoS_2 was used as an active phase and functionalized graphene sheets (GS) as the support. The support was sonicated with a solution of ammonium tetrathiomolybdate and the solid obtained was vacuum dried before further treatments using two different methods. In the first method, the solid was subjected to microwave radiation in an oven to obtain a Mi-MoS_2/GS catalyst, whereas the second method involved thermal treatment without irradiation to obtain a Th-MoS_2/GS catalyst. In addition, an MoS_2/AC catalyst was prepared and tested for comparison. The results in Figure 9.10[41] show that the microwave treatment produced a much more active catalyst than the conventional thermal treatment. In the comparison with the MoS_2/AC catalyst, the Mi-MoS_2/GS catalyst was more active at all indicated temperatures except at 200 and 240 °C, when bulk MoS_2 exhibited higher activity.

Graphene nanoribbon (GNR) prepared by unzipping of CNTs was used by Yang *et al.*[42] as the support for MoS_2 catalysts. Both single-layer (SL-MoS_2/GNR) and few-layers (FL-MoS_2/GNR) catalysts were prepared in addition to a

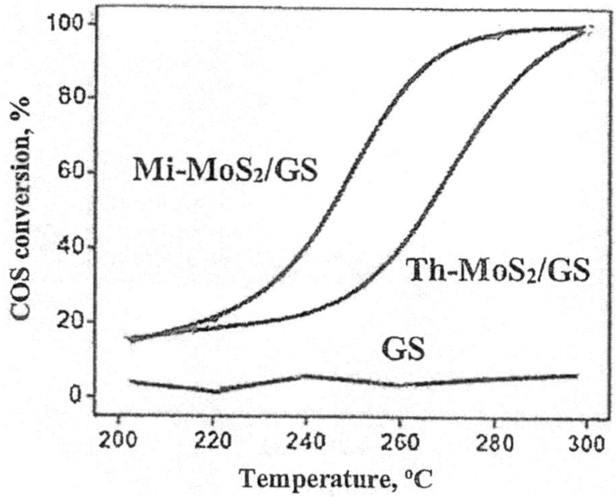

Figure 9.10 Effect of temperature and preparation method on catalyst activity. Reproduced from ref. 41 with permission from Elsevier, Copyright 2014.

pure multi-layer MoS_2 (ML-MoS_2) catalyst. The superiority of the SL-MoS_2/GNRs catalyst in the HDS of COS was confirmed. Thus, in the temperature range 180–280 °C, the following catalyst activity order was established: SL-MoS_2/GNR > FL-MoS_2/GNR > ML-MoS_2. The best performance of the SL-MoS_2/GNR catalyst resulted from a high concentration of active sites on the surface.

Figure 9.11 shows the results for the evaluation of monometallic Ni and Co and bimetallic Ni–Co catalysts in the HDS of COS in a study by Wang *et al.*[43] These catalysts were prepared by the impregnation of graphene sheets (GS) with a solution of methyl acetate. After washing and drying, the catalysts were reduced either in a flow of H_2 or by a dielectric barrier discharge (DBD) plasma in an H_2 atmosphere. The catalysts were tested in the temperature range 413–573 K in a continuous system. The reaction mixture comprised H_2–N_2 (10 : 90) mixed with COS. It is evident from Figure 9.11[43] that the DBD reduction had a more beneficial effect on catalyst activity than the reduction in H_2, particularly on bimetallic catalysts, which were more active than monometallic catalysts.

9.2.6 Summary of Results on GO- and rGO-supported Catalysts

In the various studies conducted over noble metal catalysts supported on graphene-derived supports, summarized in Table 9.12, the testing of the CALD reactant under similar conditions to those over the corresponding catalysts supported on CNTs and CNFs (Tables 7.8 and 8.10, respectively) deserves attention. First, the selectivity with respect to CALC of the catalysts

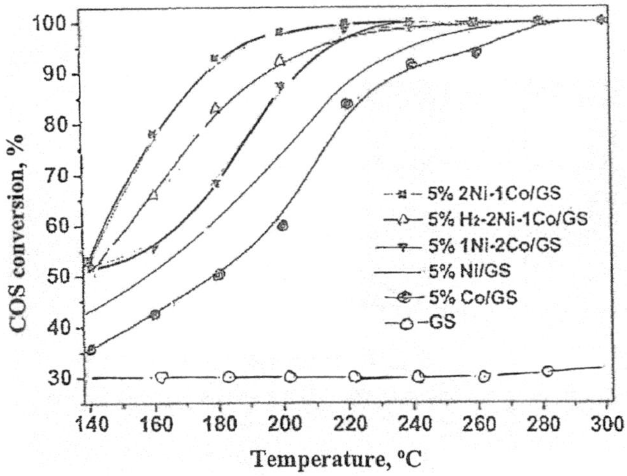

Figure 9.11 Effect of temperature and catalyst type on COS conversion. Reproduced from ref. 43 with permission from Elsevier, Copyright 2015.

supported on graphene supports was low compared with that of the latter catalysts. This suggests that the curvature of the surface of the CNT and CNF supports played a certain role in defining the geometry of the active sites. Apparently, the adsorption of CALD on the active metal *via* C=O bonds, which is a requirement for CALC formation, was more favorable on active metals supported on CNTs and CNFs than that on the flat surface available on GO and rGO supports. It is suggested that for the latter supports, the association of active metals with peripheral regions may facilitate the activated adsorption of CALD. However, the contribution of these regions to the overall surface area of the flat GO and rGO supports is rather small.

It is believed that, if present in excess, the O-containing groups on an rGO support may interfere with the activated adsorption of CALD on active sites. This is supported by the results of Ji *et al.*[7] in a study in which the rGO prepared by a novel method[8] had a much lower content of oxygen than the rGOs usually used as catalyst supports. During the HYD of CALD, the Pt catalyst supported on the novel rGO exhibited a high selectivity with respect to CALC.

The reactants derived from lignocellulosic biomass (*e.g.* LA, FF, HMF, fructose, cellobiose and vanillin) have also been tested. Studies in which GO- and rGO-supported noble metal-containing catalysts were compared with those supported on AC and oxidic supports indicated the superiority of the former catalysts. However, a more extensive database is necessary for the optimization and/or confirmation of the activity and selectivity differences. Other reactants have been evaluated to a lesser extent than CNT- and CNF-supported catalysts.

Table 9.12 Catalysts supported on graphene-derived supports.

Catalyst	Reactant/feed	Conditions	Ref.
Noble metal-containing oxygenates			
1. Pt/rGO; Pt/CNT; Pt/AC	CALD	313 K; 2 MPa; batch; 2-propanol	2
2. Pd/FLG; Pd/CNT	CALD	353 K; 1 atm H_2; batch; dioxane	4
3. Ru/O; Ru/CNT	CALD	373 K; 1 MPa H_2; batch	5
4. Pd/rGO	CALD	343 K; 2 MPa H_2; batch; aq.	6
5. Pt/graphene; Pt/CB	CALD	333 K; 1 MPa H_2; batch; 2-propanol	7, 8
6. Pt/rGO; Pt/GO; Pt/AC; Pt/CNT; Pt/SiO_2	Cellobiose	463 K; 5 MPa H_2; batch	9
7. Ru/GO; Ru/rGO; Ru/G; Ru/AC; Ru/SiO_2; Ru/TiO_2; Ru/Al_2O_3	FF	293 K; 2 MPa H_2; batch; aq.	10
8. Pt/rGO; Pt/Al_2O_3; Pt/SiO_2	HMF	393 K; 3 MPa H_2; batch; butanol	11
9. Pd–Zr/GO	Fructose	433 K; 1 MPa H_2; batch	12
10. Pd/rGO	Vanillin	373 K; 1 MPa H_2; batch; aq.	13
11. Ru/rGO; Ru/AC	LA	538 K; 1–2.5 MPa H_2; continuous; dioxane	14
12. Ru/rGO	LA	323 K; 2 MPa H_2; batch; aq.	15
13. Ru/FLG; Ru/AC; Ru/CB; Ru/SiO_2	LA	300 K; 2 MPa H_2; batch	16
Other reactants			
14. Ru/rGO; Ru/AC	Benzene; toluene	403 K; 4 MPa H_2; batch; decane	17
15. Ru/rGO; Ru/CNT	Benzene	343 K; 1 MPa H_2; batch	19
16. Pd/rGO; Pd–Ni/rGO	Chlorophenol; chlorobenzene	323 K; NH_3BH_3; batch; propanol	19, 21
17. Rh/rGO	Quinoline	353 K; 1–3 MPa; batch	20

Table 9.12 (*Continued*)

Catalyst	Reactant/feed	Conditions	Ref.
Transition metal-containing catalysts/oxygenate reactants			
18. Ni–Cu/rGO; Ni/rGO; Cu/rGO	CALD	423 K; 2 MPa; batch; methanol	25
19. Ni–Co/rGO; Ni/rGO; Co/rGO	Lignin (β-O-4)	443 K; 2 MPa H_2; batch; aq. ethanol	26–28
20. NiB/rGO; CoB/rGO; NixP/rGO; CoxP/rGO	Phenol	495 K; 4 MPa H_2; batch; dodecane	29
21. Mo_2C/rGO; Mo_2C/AC; Mo_2C/MC; Mo_2C/SC; Co–Mo/Al_2O_3	Soybean oil; oleic acid	623 K; 5 MPa; continuous	30
Petroleum feeds and reactants			
22. Ni–Mo/rGO–TiO_2	DBT	353 K; 2 MPa; continuous; tetralin	31
23. Co/N-doped G	Quinoline	393 K; 3 MPa; batch; methanol	32
24. Co–Mo/G functionalized; Co–Mo/Al_2O_3	Naphtha; diesel	573 K; 1.5 MPa; continuous	33
25. Co–Mo/GO; Co–Mo/Al_2O_3	Naphtha; diesel	588 K; 1.5 MPa H_2; continuous	34–36
26. Ni–Mo/GO; Ni–Mo/Al_2O_3	Heavy naphtha	563 K; 3 MPa H_2; continuous	37
27. Ni/rGO; Fe/rGO	Heavy feed	553 K; batch; tetralin	38–40
Conversion of COS			
28. MoS_2/GS; MoS_2/AC	COS	473–573 K; 1 atm H_2–N_2; continuous	41
29. FL-MoS_2/GNR; ML-MoS_2/GNR	COS	453–553 K	42
30. Ni/GS; Co/GS; Ni–Co/GS	COS	413–573 K; H_2–N_2; continuous	43

The data in Tables 7.33 and 8.10, involving transition metal catalysts supported on CNTs and CNFs, respectively, are much more extensive than those in Table 9.12 for the corresponding GO- and rGO-supported catalysts. For the latter catalysts, studies in which graphene-derived supports were compared with other supports are of particular importance. A better performance of rGO-supported catalysts compared with commercial HPR catalysts was noted particularly when petroleum-derived reactants and feeds were tested. This can be attributed to the beneficial effect of the graphene-derived supports on active metals compared with other conventional supports.

References

1. E. Furimsky, *Ind. Eng. Chem. Res.*, 2017, **56**, 11359–11371.
2. Z. Sun, Z. Rong, Y. Wang, Y. Xia, W. Du and Y. Wang, *Adv. RSC*, 2014, **4**, 1874–1878.
3. J. Nakamura and T. Kondo, *Top. Catal.*, 2013, **56**, 1560–1568.
4. T. Truong-Huu, K. Chizari, I. Janowska, M. S. Moldovan, O. Ersen, L. D. Nguyen, M. J. Ledoux, C. Pham-Huu and D. Begin, *Catal. Today*, 2012, **189**, 77–82.
5. Y. Wang, Z. Rong, Y. Wang and J. Qu, *J. Catal.*, 2016, **333**, 8–16.
6. R. Nie, M. Miao, W. Du, J. Shi, Y. Liu and Z. Hou, *Appl. Catal., B*, 2016, **180**, 607–613.
7. X. Ji, X. Niu, B. Li, Q. Han, F. Yuan, F. Zaera, Y. Zhu and H. Fu, *ChemCatChem*, 2014, **6**, 3246–3253.
8. L. Wang, C. Tian, H. Wang, Y. Ma, B. Wang and H. Fu, *J. Phys. Chem. C*, 2010, **114**, 8727–8733.
9. D. Wang, W. Niu, M. Tan, M. Wu, X. Zheng, Y. Li and N. Tsubaki, *ChemSusChem*, 2014, 7, 1398–1406.
10. C. Ramirez-Barria, M. Isaacs, K. Wilson, A. Guerrero-Ruiz and I. Rodríguez-Ramos, *Appl. Catal., A*, 2018, **563**, 177–184.
11. J. Shi, Y. Wang, W. Du and H. Hou, *Fuel*, 2016, **163**, 74–79.
12. R. Insyani, D. Verma, S. M. Kim and J. Kim, *Green Chem.*, 2017, **19**, 2482–2490.
13. A. A. Ibrahim, A. Lin, F. Zhang, K. M. Abou Zeid and M. S. El-Shall, *ChemCatChem*, 2017, **9**, 469–480.
14. P. P. Upare, M. Lee, S. K. Lee, J. Yoon, J. B. Bae, D. W. Hwang, U. H. Lee, J. S. Chang and Y. K. Hwang, *Catal. Today*, 2016, **265**, 174–183.
15. Y. Wang, Y. Wang, T. Wang, Q. Du, Y. Wang and J. Qu, *ACS Sustainable Chem. Eng.*, 2017, **5**, 1538–1548.
16. C. Xiao, T. W. Goh, Z. Qi, S. Goes, K. Brashler, C. Perez and W. Huang, *ACS Catal.*, 2016, **6**, 593–599.
17. K. X. Yao, X. Liu, Z. Li, C. C. Li, H. C. Zeng and Y. Han, *ChemCatChem*, 2012, **4**, 1938–1942.
18. W. Xiao, Z. Sun, S. Chen, H. Zhang, Y. Zhao, C. Huang and Z. Liu, *RSC Adv.*, 2012, **2**, 8189–8193.

19. X. Guo, C. Yu, Z. Yin, S. Sun and C. T. Seto, *ChemSusChem*, 2018, **11**, 1617–1620.
20. A. Karakulina, A. Gopakumar, Z. Fei and P. J. Dyson, *Catal. Sci. Technol.*, 2018, **8**, 5091–5097.
21. W. Chang, H. Kim, G. Y. Lee and B. J. Ahn, *Res. Chem. Intermed.*, 2016, **42**, 71–82.
22. E. Furimsky, Carbons and Carbon Supported Catalysts in Hydroprocessing, in *RSC Catalysis Series*, RSC Publishing, Cambridge, UK, 2008.
23. H. Topsoe, B. S. Clausen and F. E. Massoth, Hydrotreating Catalysts, in Catalysis Science and Technology, ed. J. Anderson and M. Boudart, 1996, vol. 11, pp. 1–310.
24. E. Furimsky, *Catal. Today*, 2013, **217**, 13–56.
25. S. S. Mohire and G. D. Yadav, *Ind. Eng. Chem. Res.*, 2018, **57**, 9083–9093.
26. B. Chen, F. Li and G. Yuan, *Catal. Lett.*, 2017, **147**, 2877–2885.
27. M. P. Pechini, *U. S. Pat.* 3,3306,97A, 1967.
28. Z. Strassberger, A. H. Alberts, M. J. Louwerse, S. Tanase and G. Rothenberg, *Green Chem.*, 2013, **15**, 768–774.
29. W. Wang, P. Liu, K. Wu, S. Tan, W. Li and W. Yang, *Green Chem.*, 2016, **18**, 984–988.
30. (a) S. K. Kim, D. Yoon, S. C. Lee and J. Kim, *ACS Catal.*, 2015, **5**, 3292–3303; (b) J. Xu, Y. Guo, T. Huang and Y. Fan, *Appl. Catal., B*, 2019, **244**, 385–395.
31. H. Wang, B. Xiao, X. Cheng, C. L. Wang, Y. Zhu, J. Zhu and X. Lu, *Appl. Catal.*, 2015, **502**, 157–165.
32. Z. Wei, Y. Chen, J. Wang, D. Su, M. Tang, S. Mao and Y. Wang, *ACS Catal.*, 2016, **6**, 5816–5822.
33. Z. Hajjar, M. Kazemeini, A. Rashidi and M. Bazmi, *Fuel*, 2016, **165**, 468–476.
34. Z. Hajjar, M. Kazemeini, A. Rashidi and M. Bazmi, *Catal. Lett.*, 2015, **145**, 1660–1672.
35. Z. Hajjar, M. Kazemeini, A. Rashidi and S. Soltanali, *J. Taiwan Inst. Chem. Eng.*, 2017, **78**, 566–575.
36. Z. Hajjar, M. Kazemeini, A. Rashidi, S. Soltanali and F. Bahadoran, *J. Anal. Appl. Pyrolysis*, 2017, **123**, 144–151.
37. B. Behnejad, M. Abdouss and A. Tavasoli, *Braz. J. Chem. Eng.*, 2018, **25**, 2528–2538.
38. L. Yuan, X. Wang, K. Zhao, H. Pan, Q. Li, J. Yang and Z. Zhang, *Pet. Sci. Technol.*, 2017, **35**, 196–200.
39. X. Xing, X. Wang, H. Pan, L. Yuan, Q. Li, M. Zhang and J. Yang, *Pet. Sci. Technol.*, 2015, **33**, 1742–1748.
40. B. B. Zhang, J. L. Song, G. Y. Yang and B. X. Han, *Chem. Sci.*, 2014, **5**, 4656–4660.
41. N. Liu, X. Wang, W. Xu, H. Hu, J. Liang and J. Qiu, *Fuel*, 2014, **119**, 163–169.
42. L. Yang, X. Wang, Y. Liu, Z. Yu, R. Li and J. Qiu, *Catal. Sci. Technol.*, 2017, **7**, 693–702.
43. W. Wang, N. Xu, Z. Liu, J. Yu and J. Qiu, *Catal. Today*, 2015, **256**, 203–208.

CHAPTER 10

Catalysts Supported on Fullerenes

A summary of methods for the preparation of fullerenes relevant to catalysis was published by Fonseca *et al.*[1] and potential applications of fullerenes as supports in catalysis were discussed by Coq *et al.*[2] In HYD and HPR applications, fullerene-supported catalysts have attracted much less attention than catalysts supported on CNT, CNF and graphene-derived supports. In HYD reactions, Osaki *et al.*[3,4] observed the formation of $C_{60}H_{18}$ in the presence of catalysts such as Ru, Rh and Ir, whereas $C_{60}H_{36}$ was produced over Pd, Pt, Co and Ni catalysts. The difference was attributed to a lower d character of metallic bonds.

Kushch *et al.*[5] compared carbon nanomaterials such as SWCNTs, MWCNTs, CNFs and fullerene black (FB) as supports for the preparation of Pt catalysts. FB is a solid residue obtained during fullerene extraction from soot. A suspension of oxidized supports was treated ultrasonically before adding an $H_2PtCl_6 \cdot xH_2O$ solution. Formic acid and sodium carbonate solution were used for reduction. The solid was filtered, washed and dried. The catalysts were applied in the HYD of decene. All carbon nanomaterial-supported catalysts were more active than a commercial Pt/AC catalyst. Among the former, Pt/FB was the most active.

Braun *et al.*[6] gave a detailed account of the preparation and properties of fullerenes. FB and toluene-extracted FB (EFB) were used as supports for the preparation of Ru/FB and Ru/EFB catalysts.[7] In addition, cathode deposit (CD) and graphite (G) were used. Ru nanoparticles were deposited on the supports by thermally induced decomposition of $Ru_3(CO)_{12}$ in boiling toluene. In parallel, a model substance such as Ru_3C_{60} and also Ru/graphite and Ru/cathode carbon were prepared for comparison. The catalysts were applied in the HYD of cyclohexenone to cyclohexanone at 310 K and near atmospheric

Catalysis Series No. 37
Carbon Nanomaterials in Hydrogenation Catalysis
By Edward Furimsky

Published by the Royal Society of Chemistry, www.rsc.org

pressure H_2 in a batch system. All catalysts were highly selective for the HYD of the C=C bond leading to cyclohexanone formation, and only trace amounts of cyclohexanol were formed. Among the catalysts, the following activity order was established: Ru/C_{60} < Ru/FB < Ru/G < Ru/EFB < Ru/CD.

Ru/C_{60} catalysts with molar ratios of Ru to C_{60} of 1 : 1, 10 : 1 and 30 : 1 were prepared by Leng *et al.*[8] and applied in the HYD of CALD (Figure 5.2) in a batch reactor at 70 °C and 2 MPa of H_2. These Ru : C_{60} molar ratios were maintained during the catalyst preparation using Ru[(1,5-cyclooctadiene)-(1,3,5-cyclooctatriene)] as Ru precursor. However, the actual Ru contents of the catalysts are given in Table 10.1.[8] In terms of TOFs, the activity of the catalysts increased with increase in Ru loading but the increase was not proportional to the amount of Ru. The product distribution in Table 10.1 (at 20% conversion) indicates good selectivity of all catalysts for the conversion of CALD to CALC.

Lashdaf *et al.*[9] prepared Ru/fullerene catalysts by the impregnation of fullerenes with solutions of ruthenium(III) acetylacetonate and bis(cyclopentadienyl)ruthenium in ethanol. After the impregnation, the samples were dried at 373 K and reduced at 537 K. In parallel, a commercial Ru(5%)/AC catalyst was reduced at 673 K in H_2. The catalysts were applied in the conversion of CALD between 333 and 393 K under 0.1–4 MPa of H_2 in an autoclave. The Ru/fullerene catalysts were selective for the HYD of the C=O bond whereas the Ru/AC catalyst exhibited high selectivity for the HYD of the C=C bond.

In addition to being used as supports alone, fullerenes can also be used as part of composite supports. For example, Coq *et al.*[10] prepared a grafted silica (SiO_2–C_{60}) support for the HYD of CALD. In another study, Anderson *et al.*[11] were able to incorporate fullerene units into the one-dimensional channels of microporous aluminophosphate. Potentially, this solid may exhibit unique properties as a catalyst support, but this still needs to be confirmed experimentally.

A novel Pt catalyst supported on fullerene-grafted-silica was used for the HYD of CALD to CALC by Coq *et al.*[10] For the preparation of the catalyst, a suspension of silica in a mixture of diethoxymethylsilane and hexane was heated at 333 K in Ar. After washing with THF, the solid obtained was suspended in dichlorobenzene containing fullerene and heated at 398 K in Ar. After filtration, the brown solid was washed with dichlorobenzene to obtain fullerene-grafted silica (SiO_2–C_{60}), which was activated in vacuum at 373 K. For Pt deposition, platinum 2,5-pentanedionate was dissolved in THF and contacted with the SiO_2–C_{60}. After removing THF, the solid was vacuum

Table 10.1 Properties of Ru/C_{60} catalysts and selectivity in the HYD of CALD.[8]

Ru : C_{60} molar ratio	Ru content/ wt%	Ru particle size/nm	TOF/ h^{-1}	CALC/%	HCALC/%	HCALD/%
1 : 1	11	nd	13	65	12	23
10 : 1	49	1.26	18	77	3	20
30 : 1	55	1.34	32	65	2	33

dried at 353 K, calcined in air at 573 K and then reduced at 623 K in a flow of H_2–N_2 containing 10% H_2 to produce a Pt/SiO_2–C_{60} catalyst containing 0.8 wt% Pt and 3.42 wt% C. The thermal stability of this catalyst in both air and H_2 was fairly high. Thus, the release of fullerene fragments was observed at 700 and 800 K; these temperatures are far above those typically used during HYD. Hence testing of this catalyst in the HYD of CALD was conducted in a batch reactor at 383 K and 4.5 MPa of H_2. High selectivity of the Pt/SiO_2–C_{60} catalyst for CALC was noted. This was attributed to a unique adsorption mode involving the C=O bond on Pt particles and the aromatic ring on fullerene.

Qin *et al.*[12] evaluated the suitability of fullerene (C_{60}) as a catalyst support in comparison with that of rGO, AC and graphite. Whereas C_{60} was used as received, the other supports were pretreated. The catalysts were prepared by sonication of the supports with a solution of ammonium molybdate in deionized water to obtain 18.8 wt% of Mo on the catalyst. After drying, the solids were carburized in a flow of H_2 to obtain either Mo_2C or $Mo_2C + MoO_2$. The latter catalysts exhibited high HIS activity whereas the Mo_2C-containing catalysts were very active for HDO. The activities of the catalysts were determined in the HDO of maize oil (a mixture of palmitic, oleic and linoleic acids) in hexane in an autoclave at 260 °C and 2 MPa of H_2 for 3 h. With respect to the effect of the carbon support, the following activity order of the catalysts was established: rGO > AC > C_{60} > graphite.

References

1. A. Fonseca, H. Richter and S. C. Emberson, *Oil Gas Sci. Technol. Rev. IFP*, 1994, **49**, 413–419.
2. B. Coq, J. M. Planeix and V. Brotonsa, *Appl. Catal.*, 1998, **173**, 175–183.
3. T. Osaki, T. Tanaka and Y. Tai, *React. Kinet. Catal. Lett.*, 2003, **79**, 319–324.
4. T. Osaki, T. Tanaka and Y. Tai, *React. Kinet. Catal. Lett.*, 2003, **78**, 217–223.
5. S. D. Kushch, N. S. Kuyunko and B. P. Tarasov, *Kinet. Catal.*, 2009, **50**, 895–898.
6. T. Braun, M. Wohlers, T. Belz, G. Nowitzke, G. Wortmann, Y. Uchida, N. Pfänder and R. Schlögl, *Catal. Lett.*, 1997, **43**, 167.
7. T. Braun, M. Wohlers, T. Belz and R. Schlögl, *Catal. Lett.*, 1997, **43**, 175.
8. F. Leng, I. C. Gerber, M. R. Axet and P. Serp, *C. R. Chim.*, 2018, **21**, 346–353.
9. M. Lashdaf, A. Hase, E. Kauppinen and A. O. I. Krause, *Catal. Lett.*, 1998, **52**, 199–204.
10. B. Coq, V. Brotons, J. M. Planeix, L. C. de Ménorval and R. Dutartre, *J. Catal.*, 1998, **176**, 358–364.
11. M. W. Anderson, J. Shi, D. A. Leigh, A. E. Moody, F. A. Wade, B. Hamilton and S. W. Carr, *J. Chem. Soc., Chem. Commun.*, 1993, 533–536.
12. Y. Qin, L. He, J. Duan, P. Chen, H. Lou, X. Zheng and H. Hong, *ChemCatChem*, 2014, **6**, 2698–2705.

CHAPTER 11

Selection of Carbon Supports

Of primary interest are amorphous carbons such as ACs and CBs in comparison with carbon nanosupports such as CNTs, CNFs and graphene-derived supports. The potential of amorphous carbons as catalyst supports has been evaluated previously.[1,2] Frequently, ACs and CBs have been used for comparison in studies involving carbon nanomaterials. Some of this information was highlighted in figures and tables in earlier chapters. In most cases, it was evident that catalysts supported on carbon nanosupports exhibited a better performance than those supported on amorphous carbons. From a practical and/or commercial point of view, the cost differential between amorphous carbons and carbon nanomaterials may be decisive. This issue is discussed in Chapter 12.

In numerous studies, catalysts supported on ACs and CBs exhibited higher activity and stability than the corresponding catalysts supported on oxidic supports, particularly in the presence of water. For example, Nie *et al.*[3] compared AC and CB supports with oxidic supports such as MgO, SiO_2 and ZrO_2 as part of the Ni catalysts used in the HYD of vanillin in an aqueous phase. In this case, a commercial AC and a nitrogen-doped CB (NCB) were used for catalyst preparation. The CB was preoxidized with H_2O_2, then deposited with polypyrrole and heated under N_2 at either 600 or 900 °C to obtain NCB-600 and NCB-900 supports, respectively. These supports and undoped CB were impregnated using an aqueous solution of Ni nitrate to obtain Ni/NCB-600, Ni/NCB-900 and Ni/CB catalysts, respectively. For comparison, Ni/AC, Ni/MgO, Ni/SiO_2 and Ni/ZrO_2 catalysts, all containing 10 wt% Ni, were also prepared using the same method. The HYD of vanillin was performed in an autoclave at 150 °C and 5 MPa for 2 h. The results in Table 11.1[3] confirm the much higher activity of the Ni/NCB catalyst compared with the Ni/CB catalyst. The activity of the Ni/AC catalyst was lower than that of the Ni/CB catalysts, but it was still higher than the activity of Ni catalysts supported on oxidic supports. This indicates that under

Catalysis Series No. 37
Carbon Nanomaterials in Hydrogenation Catalysis
By Edward Furimsky

Published by the Royal Society of Chemistry, www.rsc.org

Table 11.1 Aqueous-phase HDO of vanillin over Ni catalysts.[3]

Catalyst[a]	Conversion/%	Selectivity/% HMP	Selectivity/% MMP	MMP yield/%
None	1.2	—	—	—
Ni/NCB-600	52.9	39.4	60.6	32.1
Ni/NCB-900	74.4	35.4	64.6	48.1
Ni/CB	37.8	49.4	50.6	19.1
Ni/AC	35.9	75.2	24.8	8.9
Ni/SiO_2	0.3	—	—	—
Ni/MgO	19.0	1.4	98.6	18.7
Ni/ZrO_2	3.7	79.5	20.5	0.8
2nd re-use[b]	78.1	34.9	65.1	50.8
3rd re-use	72.3	32.5	67.5	48.8
4th re-use	71.1	35.5	64.5	45.9
5th re-use	68.7	37.2	62.8	43.1

[a]Conditions: 228 mg vanillin, 20 mg catalyst, 10 mL H_2O, 0.5 MPa H_2, 150 °C, 2 h.
[b]Re-use of Ni/NCB-900.

aqueous conditions, the catalysts supported on oxidic supports may have limited applications. The most active catalyst, Ni/NCB-900, was recycled five times, after which most of the activity and selectivity of the original catalyst were retained. Other results supporting higher activity of catalysts supported on amorphous carbons in comparison with oxidic supports were given in Tables 9.5 and 9.6. This information confirmed that under aqueous conditions, catalysts supported on amorphous carbons outperformed the corresponding catalysts supported on oxidic supports.

Numerous results (Tables 9.5, 9.6 and 9.11) indicate a much better performance of catalysts supported on carbon nanosupports such as CNTs, CNFs and rGO than that of catalysts supported on AC, CB and oxidic supports. Then, in line with the results of Nie *et al.*[3] discussed above, catalysts supported on carbon nanosupports should exhibit a much better performance than those supported on oxidic supports. Some studies that demonstrate the advantages of the former catalysts are highlighted in the following discussion, and the numerous studies confirming these advantages can be identified in the earlier summary tables.

11.1 Carbon Nanotubes *Versus* Amorphous Carbons

The frequently investigated reactant CALD was tested over Ru–Pt catalysts supported on CNTs, CNFs and ACs. In rather delicate reactions such as the HYD of CALD to CALC, the performance of the AC-supported catalysts was inferior to that of the CNT- and CNF-supported catalyst, as shown in Table 7.5. Table 7.10 indicates a similar advantage of a CNT support compared with an AC support in the conversion of cellulose to sorbitol over Ru–Ni/CNT and Ru–Ni/AC catalysts. The activity and selectivity difference in favor of the CNT-supported catalyst was even more evident when an Ru/CNT catalyst was compared with Ru/Al_2O_3 and Ru/SiO_2 catalysts in the conversion of glucose to sorbitol (Table 7.11). Important results on the effect of different supports in the conversion of LA to GVL were presented in Table 7.17. The best performance of the CNT-supported catalysts was clear for all active metals, *i.e.* Ir, Ru and Pd. The corresponding AC-supported catalysts were less active. All six oxidic supports in Table 7.17 were inferior even in comparison with the AC-supported catalysts. An order of magnitude higher activity over Ru/CNT catalyst compared with that over Ru/AC catalyst was observed in the conversion of HMFF to DMF (Figure 7.5). The following activity and HDO selectivity order was observed in the conversion of guaiacol: Ru/CNT > Ru/AC > Ru/carbon aerogel (CARF) > Ru/CB (Vulcan carbon) > Ru/graphite (Figure 7.6). Thus, although different reactants and conditions were employed, the Ru/CNT catalyst outperformed other Ru catalysts.

Jung and co-workers[4,5] compared CB (Vulcan carbon) with AC and MWCNTs as supports for Ru–Re catalysts used in the HDO of guaiacol and eugenol. Under similar conditions, CB (Vulcan carbon) outperformed AC and was comparable to MWCNTs. However, the selectivity of Ru–Re/MWCNTs with respect to hydrocarbons was greater than that of Ru–Re/CB. This was attributed to a greater surface concentration of Re species on the former catalyst. It was noted that the surface areas of the Ru–Re/AC, Ru–Re/MWCNT and Ru–Re/CB catalysts were 366, 215 and 109 $m^2 g^{-1}$ and the active metal distributions were 30.6, 8.6 and 11.4 %, respectively. This suggests that in determining the efficiency of active sites, electronic effects of carbon supports on active metals played a much more important role than the textural properties of the supports.

Aromatic and unsaturated hydrocarbons have been used as model reactants to evaluate the effects of carbon nanosupports and amorphous carbons. Figures 7.10 and 7.11 compare the activities of Pt(1%)/CNT and Pt/AC catalysts in the HYD of toluene and tetralin, respectively. For the AC-supported catalysts, more than twice the amount of Pt was necessary to approach the activity of the Pt(1%)/CNT catalyst. A much better performance of a Pd catalyst supported on CNTs than that of a Pd/AC catalyst was also observed in the HYD of benzene (Figure 7.12). Figures 7.18–7.20, 7.22 and 7.23 suggest that with respect to the HYD of benzene, CNTs provide the best support regardless of the type of active metal and/or active phase. A Pd/CNT catalyst outperformed Pd/CB and Pd/AC catalysts in the selective HYD of phenylacetylene to styrene, as shown in Figure 7.16.

11.2 Carbon Nanofibers *Versus* Amorphous Carbons

Herringbone and platelet CNFs were compared with CNTs, ACs and Al_2O_3 as supports for Pt catalysts.[4,5] The catalysts were used in the conversion of CALD to CALC. The effect of conversion on selectivity is shown in Figure 8.2. The Pt/H-CNF (herringbone) catalyst outperformed all other catalysts, including Pt/CNT catalysts. The relatively high activity of the Pt/AC1 catalyst compared with the commercial Pt/AC2 catalyst may be attributed to the colloidal microwave process used for the addition of Pt to AC1. The better performance of the Pt/CNF catalyst than that of the Pt/AC catalyst was also confirmed in the conversion of citral (Figure 5.3) to geraniol and nerol. In a similar comparison of Pd catalysts in the conversion of CALD, but with a high selectivity with respect to HCALD rather than to CALC (Figure 8.3), the performance of a Pd/CNT catalyst was better than that of a Pd/CNF catalyst. As usual, the activity and selectivity of the Pd/AC catalyst were lower. The conversion of sorbitol to glycols was another reaction used to compare the effect of CNFs and ACs on the activity of catalysts.[6] In this case, Ru was used as the active metal. The higher activity and selectivity of an Ru/CNF catalyst than those of an Ru/AC catalyst were confirmed by the results in Table 8.2 for the conversion of sorbitol. When normalized to a unit of active metal, the activity of the Ru/CNF catalyst in the HYD of LA to GVL was at least an order of magnitude higher than that of the Ru/AC catalyst.[7]

Among delicate reactions, the selective HYD of alkynes to alkenes, with minimal conversion to alkanes, is of industrial importance. Table 8.3 gives the results for the selective conversion of octyne to octene. The Pd/AC catalyst exhibited the highest activity but its selectivity was very low, as indicated by the high yield of the unwanted octane. The Pd/CNF catalyst was much more suitable. The good performance of the Lindlar (Pd/$CaCO_3$) catalyst should be noted.

Chlorohydrocarbons have attracted attention in the context of waste disposal, where their conversion to benign hydrocarbons is important to minimize their environmental impact. The following activity order of Pd catalysts was established: Pd/CNF > Pd/AC ≫ Pd/graphite.[8] However, in the HDCl of chlorophenol over Pd catalysts supported on graphite, CBs and CNFs, the highest activity was exhibited by the Pd/CB catalyst.[9]

Among transition metal-containing catalysts, an Ni/CNF catalyst was much more active and selective than Ni/AC and $NiAl_2O_3$ catalysts in the conversion of cellulose to sorbitol.[10,11] This was also confirmed in the conversion of CALD to HCALC over Ni/CNF and Ni/AC catalysts, as shown in Table 8.4. Tables 8.5 and 8.6 give the results for the conversion of guaiacol and HDO of methyl palmitate, respectively, over Mo_2C catalysts supported on CNFs, CNTs and ACs. The results are not consistent enough to be able to select the best catalyst with respect to its activity and selectivity.

11.3 Graphene Supports *Versus* Amorphous Carbons

The activities and selectivities of Pt/rGO, Pt/CNT and Pt/AC catalysts were compared in the conversion of CALD with the objective of maximizing the yield of CALC.[12] Table 8.10 shows that in this regard, the Pt/rGO catalyst exhibited the best performance. The same reactant was used to compare Pd(5%) catalyst supported on few-layers graphene (FLG) with that supported on CNTs.[13] In this case, selectivity with respect to HCALD rather than to CALC was the objective. Under the same conditions, the Pd/FLG catalyst outperformed the Pd/CNT catalyst. As Table 9.1 shows, rGO was a better support than CNTs as part of Ru(4%) catalysts. Thus, at a similar selectivity, the activity of the former catalyst was more than twice that of the Ru/CNT catalyst.

Reactants of lignocellulosic origin have frequently been used for the evaluation of the effects of supports on catalyst performance. For example, the effects of graphite, AC, GO, rGO, CNTs and SiO_2 supports were evaluated in the depolymerization of cellulose over Pt catalysts. The following order of selectivity with respect to sorbitol was established: Pt/rGO > Pt/CNT > Pt nanoparticles > Pt/AC > Pt/SiO_2 > Pt/graphite.[14] Ru(4%) catalysts supported on rGO, AC, SiO_2, TiO_2, Al_2O_3 and high surface area graphite (G) supports were compared in the conversion of furfural (FF) to furfuryl alcohol (FFA). The results in Table 9.5 indicate the superiority of the rGO-supported catalysts. Table 9.6 provides further confirmation of the superiority of rGO-supported catalysts in the conversion of 5-hydroxymethylfurfural to 2,5-dimethylfuran. Based on these results, the following order of activity of the catalysts can be established: Pt/rGO > Pr/AC > Pt/SiO_2 > Pt/Al_2O_3. The Ru/rGO catalyst exhibited a much better performance than the Ru/AC catalyst in the conversion of LA to GVL.[15] In another study, Ru catalysts supported on FLG, AC and CB supports and a commercial Ru/AC catalyst were compared in the conversion of LA to GVL.[16] It was evident that the best performance was shown by the Ru/FLG catalyst.

The study published by Kim *et al.*[17] is one of the few in which an active phase comprising a conventional metal was investigated. In this study, an Mo_2C/rGO catalyst was used for the HDO of oleic acid and soybean oil with the corresponding catalysts supported on AC, a spherical carbon (SC) and mesoporous carbon (MC) and also a sulfided Co–Mo/Al_2O_3 catalyst. As Figure 9.6 shows, the Mo_2C/rGO catalyst outperformed the other catalysts. Of particular importance are the results in Figure 9.7, showing a much higher stability of the Mo_2C(17%)/rGO catalyst compared with that of the conventional sulfided Co–Mo/Al_2O_3 catalyst. The superiority of the Co–Mo catalyst supported on graphene in comparison with the conventional Co–Mo/Al_2O_3 catalyst was also observed in the HDS of a petroleum feed, as indicated in Table 9.11 and Figure 9.9. A similar observation was also made for a conventional Ni–Mo/Al_2O_3 catalyst.[18]

11.4 Selection of Carbon Nanosupports

The database necessary for selecting the best supports among carbon nanosupports such as CNTs, CNFs, CNHs and graphene-derived supports is not available. In a few studies, not more than two carbon nanosupports were compared under similar conditions. Table 7.5 may suggest that during the conversion of CALD, catalysts supported on a CNF support were better than those supported on CNTs. However, it is believed that additional testing is needed to verify these results. In a study on the HDO of guaiacol,[19] an Mo_2C catalyst supported on CNTs outperformed the catalyst supported on CNFs, as shown in Table 8.6. Little difference in the activity and selectivity of Mo_2C

catalysts supported on CNTs and CNFs was observed in the HDO of methyl palmitate (Table 8.7).

References

1. E. Furimsky, *Carbons and Carbon Supported Catalysts in Hydroprocessing*, RSC Publishing, Cambridge, UK, 2008.
2. P. Serp, M. Corrias and P. Kalck, *Appl. Catal.*, 2003, **253**, 337–358.
3. R. Nie, H. Yang, H. Zhang, X. Yu, X. Lu, D. Zhou and Q. Xia, *Green Chem.*, 2017, **19**, 3126.
4. K. B. Jung, J. Lee, J.-M. Ha, H. Lee, D. J. Suh, C.-H. Jun and J. Jae, *Catal. Today*, 2018, **303**, 191–199.
5. A. Jung, A. Jess, T. Schubert and W. Schutz, *Appl. Catal.*, 2009, **362**, 95–105.
6. L. Zhao, J. H. Zhou, Z. J. Sui and X. G. Zhou, *Chem. Eng. Sci.*, 2010, **65**, 30–36.
7. Y. Yang, C.-J. Sun, D. E. Brown, L. Zhang, F. Yang, H. Zhao, Y. Wang, X. Ma, X. Zhang and Y. Ren, *Green Chem.*, 2016, **18**, 3558.
8. C. Amorim, G. Yuan, P. M. Patterson and M. A. Keane, *J. Catal.*, 2005, **234**, 268–281.
9. J. A. Baeza, L. Calvo, N. Alonso-Morales, F. Heras, S. Eser, J. J. Rodriguez and M. A. Gilarranz, *Chem. Eng. Sci.*, 2018, **176**, 400–408.
10. S. Van de Vyver, J. Geboers, M. Dusselier, H. Schepers, T. Vosch, L. Zhang, G. Van Tendeloo, P. A. Jacobs and B. F. Sels, *ChemSusChem*, 2010, **3**, 698–701.
11. S. Van de Vyver, J. Geboers, W. Schutyser, M. Dusselier, P. Eloy, E. Dornez, J. W. Seo, C. M. Courtin, E. M. Gaigneaux, P. A. Jacobs and B. F. Sels, *ChemSusChem*, 2012, **2**, 1549–1558.
12. Z. Sun, Z. Rong, Y. Wang, Y. Xia, W. Du and Y. Wang, *Adv. RSC*, 2014, **4**, 1874–1878.
13. T. Truong-Huu, K. Chizari, I. Janowska, M. S. Moldovan, O. Ersen, L. D. Nguyen, M. J. Ledoux, C. Pham-Huu and D. Begin, *Catal. Today*, 2012, **189**, 77–82.
14. B. K. Pradhan, G. U. Sumanasekera, K. W. Adu, H. E. Romero, K. A. Williams and P. C. Eklund, *Physica B*, 2002, **323**, 115–121.
15. P. P. Upare, M. Lee, S. K. Lee, J. Yoon, J. B. Bae, D. W. Hwang, U. H. Lee, J. S. Chang and Y. K. Hwang, *Catal. Today*, 2016, **265**, 174–183.
16. C. Xiao, T. W. Goh, Z. Qi, S. Goes, K. Brashler, C. Perez and W. Huang, *ACS Catal.*, 2016, **6**, 593–599.
17. S. K. Kim, D. Yoon, S. C. Lee and J. Kim, *ACS Catal.*, 2015, **5**, 3292–3303.
18. B. Behnejad, M. Abdouss and A. Tavasoli, *Braz. J. Chem. Eng.*, 2018, **25**, 2528–2538.
19. E. Santillan-Jimenez, M. Perdu, R. Pace, T. Morgan and M. Crocker, *Catalysts*, 2015, **5**, 424–441.

CHAPTER 12

Future Perspectives

In various HYD and HPR reactions, catalysts containing noble metals and transition metals supported on carbon nanosupports (*e.g.* CNTs, CNFs, CNHs, GO and rGO) outperformed the corresponding catalysts supported on conventional oxidic supports (γ-Al_2O_3, SiO_2, *etc.*) and also on amorphous carbon supports (*e.g.* CBs and ACs). Problems associated with the low reactivity and hydrophobicity of pristine carbon nanomaterials can be alleviated by introducing irregularities into their structure. For example, doping with different elements (*e.g.* N, S, P, B) results in the replacement of some of the carbon atoms in the graphene structure of carbon nanomaterials. In addition, several methods have been developed for the functionalization of pristine carbon nanomaterials. Using these methods, various functional groups (–OH, –COOH, –SO_3H, –NH_2, *etc.*) can be introduced into the pristine carbon nanomaterials. Consequently, the reactivity of the carbon nanomaterials is significantly enhanced. The originally neutral carbon supports can be made acidic by introducing acidic groups capable of donating protons. With such supports, bifunctional catalysts containing various active metals can be prepared.

The electronic effects induced by the graphene structure of the carbon nanosupports on active metals ensured the efficient distribution of the metals. In this regard, the type of doping and functionalization of carbon supports may play a critical role. Catalysts with a particle size of the active metals ranging from a few nanometers up to more than 100 nm could be prepared. The nanosize of active metals is one of the parameters responsible for the remarkable performance of catalysts supported on carbon nanomaterials. The activity and selectivity of catalysts can be influenced by preparing them with the optimal size of active metal particles. Experimental procedures have been developed for the preparation of catalysts supported on carbon nanosupports suitable for a specific reaction.

Catalysis Series No. 37
Carbon Nanomaterials in Hydrogenation Catalysis
By Edward Furimsky

Published by the Royal Society of Chemistry, www.rsc.org

The numerous reactants and feeds that have been evaluated over carbon nanomaterial-supported catalysts include oxygenates such as reactants with α,β-conjugated C=C and C=O double bonds (*e.g.* CALD and citral), reactants derived from lignocellulosic components such as cellulose, hemicellulose and lignin, and fatty acids and esters derived from triglycerides present in algae and vegetable oil biomass. For the oxygenated reactants and feeds, the catalyst evaluation was conducted under aqueous conditions and also in the presence of reaction water only. Because of the hydrophobicity of carbon supports, high stability of the catalysts under such conditions has been observed. Thus, little decline in catalyst performance was noted after several utilization–regeneration cycles. In addition to reactants and feeds of petroleum origin, hydrocarbons such as alkenes, alkynes, dienes and aromatics have been studied as part of the efforts to develop catalysts for the selective HYD of hydrocarbon streams before further processing. For example, traces of acetylene and phenylacetylene must be removed from ethylene and styrene streams, respectively, before polymer preparation. For environmental and safety reasons, the HDCl of chlorinated reactants has been studied. In most of these applications, catalysts supported on carbon nanomaterials outperformed the corresponding catalysts supported on oxidic supports and also on amorphous carbon supports. As a general observation, noble metal-containing catalysts supported on carbon nanosupports exhibited higher activity than the corresponding catalysts containing transition metals. This was indicated by the milder conditions required over the former catalysts to achieve a similar reactant conversion.

The unique selectivity observed in the HYD of reactants with α,β-conjugated C=C and C=O double bonds over catalysts supported on carbon nanomaterials has been noted. Thus, a high rate of the HYD of C=O bonds relative to that of C=C bonds can be selectively achieved in spite of the thermodynamic predictions favoring the HYD of C=C bonds. This results from the structure of the transition state comprising the reactant and active site. In this case, coordination of active metals with C=O bonds rather than C=C bonds is decisive. Then, the flow of active hydrogen from the active metal to the C=O bond, leading to the formation of a C–OH entity, is facilitated more readily. Under more severe conditions (*e.g.* during HPR), *i.e.* when complete HDO is the objective, the length of the C=O bond is extended until it is cleaved. Subsequently, oxygen remains associated with the active metals, as indicated by DFT calculations. Such an association deactivates the catalyst. To prevent catalyst deactivation, the use of a high pressure of H_2 at relatively low temperatures has been noted.

Among carbon nanosupports, most attention has been paid to catalysts supported on CNTs. In this case, noble metal-containing catalysts could be tailor-made to achieve a high conversion of various reactants with high selectivity with respect to the products of interest. Thus, an extensive database is now available on the conversion of CALD and citral and also that of monomeric reactants derived from lignocellulosic biomass and fatty acids obtained from algae and vegetable oil biomass. In a similar fashion, although to a lesser extent, catalysts supported on CNFs have been studied. The potential of CNHs as supports for HYD and HPR catalysts has not yet been fully realized. Graphene materials such as GO and rGO have been evaluated as catalyst supports in comparison with pristine graphene. In several studies, catalysts supported on carbon nanosupports were compared with the corresponding catalysts supported on amorphous carbons such as ACs and CBs. In most cases, the former catalysts exhibited higher activity and selectivity with respect to desirable products.

Studies indicating a remarkable performance of catalysts supported on carbon nanosupports in comparison with conventional catalysts that have been used commercially can be identified. For example, a useful conversion of some reactants can be achieved at ambient temperature with near atmospheric pressure H_2. However, there is little evidence to suggest that the former catalysts are being used on an industrial scale. This may be attributed to several factors that require additional clarification. First, the cost of the production of carbon nanosupports on an industrial scale can now be determined when the experimental methodology, including necessary pretreatments of the carbon supports, have been well described. Similarly, several methods for the preparation and post-treatment of catalysts have been described in detail. This allows the selection of the optimal technique for catalyst preparation.

Overwhelming evidence has confirmed the advantages of carbon nanomaterials such as CNTs, CNFs, GO and rGO as supports of catalysts for HYD and HPR applications compared with amorphous carbons such as ACs and CBs and the traditionally used oxidic supports. This has been confirmed for catalysts containing noble metals and transition metals as active metals. In spite of this evidence, there is little information to suggest that carbon nanosupports are being used as supports for the catalysts used in commercial applications. This can be attributed to a significant cost differential of the carbon nanosupports relative to other supports.

Table 12.1 summarizes some prices of carbon nanomaterials, amorphous carbons, graphite and traditional oxidic supports such as Al_2O_3 and SiO_2. These data can be readily accessed on-line.[1–4] In view of the frequent fluctuations in commodity prices, these values should be considered as an indication of relative price ranges among the listed solids rather than stable prices. It is emphasized that the prices of carbon nanomaterials in Table 12.1 (*e.g.* CNTs, CNFs, GO and rGO) are given per gram whereas the prices of the amorphous carbon solids (*e.g.* graphite, AC and CB) are given per ton. This comparison indicates price differences of several orders of magnitude between these two groups of carbon solids. The prices of the most frequently used oxidic supports, Al_2O_3 and SiO_2, are in the range of those of amorphous carbons. Therefore, the significantly higher cost of carbon nanomaterials may be the main reason why these supports are hardly utilized in commercial catalysis in spite of their remarkable effects on activity and selectivity in various HYD and HPR reactions.

Table 12.1 Costs of carbon nanomaterials and amorphous carbons.[1–4]

Material	Price
Carbon nanomaterials	*US$ per gram*
GO	170–225
rGO	400–450
CNTs	0.7–15
CNTs[a]	1–30
CNFs	65
	589 per 25 g
C_{60}	35–40
C_{60} (99.9)	225–250
Graphite	*US$ per ton*
Graphite	700–1500
Amorphous carbons	*US$ per ton*
AC	200–2000
CB	500–2000
Oxidic supports	*US$ per ton*
Al_2O_3	300–500
SiO_2	200–4000

[a]Functionalized CNTs.

The potential for the commercial utilization of carbon nanomaterials as supports for HYD catalysts is much greater than that for the catalysts used in HPR applications. Thus, practical experience has shown that large quantities of catalysts are required for the production of fuels using any HPR process. These quantities could be several orders of magnitude greater than the amounts of catalyst used for the production of some fine chemicals. Hence the high activity of catalysts supported on carbon nanosupports combined with a high selectivity for the products of interest may favor these supports in spite of the high cost.

It is believed that a price comparison of carbon nanomaterials with other solids was desirable to put the commercial use of the former into perspective. Obviously, developments in the methodology for the production of carbon nanomaterials with the aim of decreasing their cost would be of primary importance. Then the unit price would be expected to decrease with increasing capacity of production. In this regard, little information is currently available in the literature to be able to draw any conclusions.

References

1. https://www.cheaptubes.com/cheap-tubes-inc-online-shop/.
2. https://www.alibaba.com/showroom/activated-carbon-price.html.
3. https://www.focus-economics.com/commodities/base-metals/alumina.
4. https://www.alibaba.com/showroom/silica-prices.html.

Subject Index